Trees

Trees

From Root to Leaf

With over 450 illustrations

Paul Smith

Foreword by Robert Macfarlane

Contents

Fruits 206

Symbiosis 238

Trees and Us 270

Foreword
by Robert Macfarlane

Trees are among the central characters in the oldest surviving work of world literature. *The Epic of Gilgamesh* – first set down in cuneiform script upon clay tablets around 2100 BCE in Ancient Mesopotamia – tells of how the god-king Gilgamesh and the wild-man Enkidu travel on foot across seven mountain ranges to reach the Sacred Cedar Wood. The Cedar Wood, before they reach it, is a place of tranquil splendour. There the great trees have grown undisturbed to immense heights; monkeys haunt their branches, and birdsong fills the air. Gilgamesh and Enkidu, however, approach the wood not as a site of wonder but as a site of plunder. First, they brutally slay the forest's guardian spirit, Humbaba. Then they set about the cedars with their axes, felling several of the mightiest trees to make a raft and to fashion a temple door. Their actions not only cause the downfall of the forest, but also lead to Enkidu's death, which the gods mete out as punishment for the damage he and Gilgamesh have caused.

The Epic of Gilgamesh contains, we might say, the first of the tellings of the first of the fellings: the earliest act of ecocide. So much of humanity's complex relationship with trees is embedded in the story: trees as cultural inspiration, trees as remarkable living beings with which we share a fragile Earth, trees as the generous makers of biotic communities – but also trees treated wantonly as subject and commodity, and the calamities that can follow when trees and forests are flagrantly misused.

All of these themes recur in Paul Smith's magisterial *Trees*, more than four millennia on from *Gilgamesh*. They run through the book you are about to read like heartwood. Smith is a plant ecologist by training and profession, and his botanical knowledge is unsurprisingly profound. His account of the role of trees on Earth – 'our muses, protectors and silent companions', as he puts it – is also, though, founded upon deep historical and cultural roots. Richard Powers, on the second page of his extraordinary arboreal novel of 2018, *The Overstory*, imagines trees speaking to their human neighbours and co-citizens: 'If only your mind were a greener thing', they whisper, 'we would drown you in meaning.' Smith's book helps make its reader's mind a 'greener thing'.

Trees appears at a moment when arboreal life on the planet is arguably at its most imperilled since trees first emerged, around 350 million years ago. In 2021, the 'State of the World's Trees' report was published, offering the first global list of tree species and their respective levels of endangerment. Out of a worldwide total of around 60,000 tree species, c. 17,500 – around 30 per cent – were found to be threatened with extinction, and 440 to be on the brink of disappearance. Forest clearance for logging, crops, grazing and development was the main drivers of this decline. Even as we suffer through our own pandemic, it is worth remembering that trees are also plagued by their own terrible contagious diseases. Ash dieback (*Hymenoscyphus fraxineus*) is expected to kill around 80 per cent of the UK's ash trees in the next three decades. I have walked through whole hillsides in the South Downs where the ashes stand grey and dying, and I have heard the tell-tale death-rattle of their brittle branches knocking against one another when the wind rises. The mountain pine beetle, its population no longer checked by cold winters, is devastating conifer forests across the United States. Meanwhile, species such as the Douglas fir or the canyon live oak are on the march as climate refugees, their populations pushed out of long-held territories by rising temperatures and wildfires.

In 1658, the early modern physician and scholar Sir Thomas Browne coined a beautiful verb: 'to interarborate'. Interarboration, in Browne's account, is the lacing and weaving together of branches. We, as humans, are ourselves inter-arborated – drastically interdependent with trees and forests. Smith shows us here how trees have grown into and through almost every aspect of our existence, from furniture to architecture, urban planning to textiles, food and drink. Street trees – those static superheroes, those tireless ecosystem-service providers – offer us shade, reduce surface temperatures in urban heat islands, clean city air of its pollutants, store carbon, join pockets of habitat and support communities of creaturely life. The goods with which trees furnish us are not only material, of course: they make metaphor as well as glucose. We have forests inside our minds, for instance: nano-scale imagery shows that the structures of human nerve cells closely resemble the spreading canopies of certain trees, and neuroscientists have called these branching projections 'dendrites', from the Greek word 'dendron', meaning 'tree'. Where they overlap (interarborate) they are said to form a 'dendritic arbor'. We think with trees.

In my experience, to walk in a wood is to take issue with Socrates's declaration that 'Trees and open country cannot teach me anything, whereas men in towns do.' Time is kept and curated in different ways by trees, and it can be experienced in different ways when one is among them. It is beyond our capacity to comprehend that the American hardwood forest waited 70 million years for people to come and live in it – but the effort at comprehension is itself worthwhile. It is valuable and disturbing to know that big oak trees can take 300 years to grow, 300 years to live and 300 years to die.

Gauri Ragini
c. 1650
Folio from a *Ragamala*.
Opaque watercolour and gold on paper (painting).

Page 6: The Tree Collector
Rachel Campbell, 2019
Oil on linen.

Such knowledge, seriously considered, changes the grain of the mind. My favourite genus of city tree is certainly the magnolia, with its gorgeous goblet-like flowers. The carpels of magnolias are unusually tough: it is theorized that this is because magnolias evolved in the late Cretaceous, around 90 million years ago. This was before bees existed, and so their carpels needed to be able to resist damage by the chitinous feet of the beetles that were their only pollinators. Seeing a suburban magnolia always gives me a thrilling burst of Earth history: dinosaurs once browsed among these trees!

William Blake famously observed that 'the tree that moves some to tears of joy is in the eyes of others only a green thing that stands in the way'. The inexorable march of the technocratic metaphysic has widely reduced perception of the arboreal world to timber, lumber or obstacle. Wonder has been subdued by instrumentalism. One of the great contemporary challenges is now how to retrieve and popularize an almost-animist ontology, in which the lively, miraculous being-hood of species other than our own is recognized and respected. The structure of Smith's work is a contribution to such a project. It organizes itself in terms of the seven wonders – Seed, Leaf, Form, Bark, Wood, Flower, Fruit – that together conjure the magic of Tree. His book offers us a reminder that, as he puts it, 'ultimately, we are all symbionts with nature.' Or, in the unforgettable phrase of Ursula K. Le Guin, 'the word for world is forest'.

> The goods with which trees furnish us are not only material, of course: they make metaphor as well as glucose.

Foreword by Robert Macfarlane

Introduction

Trees are among the largest living things on Earth. They cover around a third of the world's land surface and play a major role in our environmental systems – influencing everything from the water, nutrient and carbon cycles to the global climate. Forests provide a home

for a wide variety of other organisms, supporting at least half of the Earth's terrestrial plant and animal species, but trees flourish in a tremendous range of environments across the globe.

Bristlecone pine
Pinus longaeva
The bristlecone pine can survive for more than 4,500 years, making it the world's longest-lived tree species. It is found in the mountains of Utah, Nevada and eastern California.

Trees have forged their roots in woodlands, shrublands, grasslands, coastal and rocky ecosystems, deserts, savannahs and wetlands, as well as being planted by humans in artificial and urban environments from London to Singapore. It is the extraordinary diversity of trees – vital to our human lives, to our planet and as a source of inspiration to the people and the cultures they build – that this book celebrates.

Trees can reach great heights of over 100 metres (350 feet) and can weigh more than 1,000 metric tonnes (985 imperial tons) – that's more than six times the weight of a blue whale.

The chapters that follow chronicle the various and colourful stages of the tree life-cycle, from seeds and leaves to fruit, flowers and bark. Although there is no universally accepted definition of a tree, the characteristic that separates trees from other plants is a woody stem or trunk that lives for many years [pg.83].

Trees can reach great heights of over 100 metres (350 feet) and can weigh more than 1,000 metric tonnes (985 imperial tons) – that's more than six times the weight of a blue whale. In spite of their importance, their ubiquity, their cultural significance and utility to humans, it was only in 2017 that the first complete list of the world's tree species was published. The current total stands at a staggering 58,497 species, a number that is constantly revised as new species are added or names and classifications are changed.

For a new tree species to be described and accepted by scientists, a botanist must find the tree in the wild, collect specimens of leaves, flowers and fruits, bring them back to a herbarium (a library of pressed plant specimens) and compare them with the samples of similar species. If they think they have found a new species, the botanist will publish a description in a scientific journal for peer review. Even using this rigorous and disciplined approach, there has been plenty of room for duplication of effort and error in the field, resulting in the average tree species being given three or four different scientific names over the centuries. It is only in recent years – thanks to the information-technology revolution and advances in molecular biology – that botanists have been able to rapidly exchange digital and genetic information about specimens to work out exactly what is what. With this background, it's hardly surprising that it has taken so long to describe the dazzling array of tree species on our planet.

It is in the tropics, with their carpets of dense rainforest, that species diversity really takes off. Brazil has the most tree species in the world (8,715), followed by Colombia (5,776) and Indonesia (5,142). Nearly 60 per cent of all tree species are endemic to a single country, meaning that there are lots of rare trees out there. Although many tree species are rare, they are all vitally important.

For modern humans, trees continue to provide shelter and food together with a wide range of products of huge value to local livelihoods, national economies and global trade.

Further north, tree diversity tends to diminish, but there's strength in numbers. Apart from the Arctic and the Antarctic (which has no trees), the region with the fewest tree species is the Nearctic region of North America, with fewer than 1,400 species. In fact, although boreal regions across the northern latitudes are not particularly diverse, they more than make up for this in the sheer volume of trees, with temperate forests comprising about half of all global forest biomass.

Wherever trees are found, they are crucial to the environment around them, supporting a wide range of other species from their position at the start of innumerable food chains and the base of ecological trophic pyramids. Despite their importance, the recently published *Global Tree Assessment* shows that more than 17,500 tree species (some 30 per cent of the total figure) are currently threatened with extinction. This has the potential to lead to an extinction cascade and the collapse of entire ecosystems. Unfortunately, our global economic system and the political structures that maintain it have paid little attention to the value of 'natural capital' upon which all life depends. This is not only a tragedy for the tree species concerned but also for the myriad species that depend on them, including us.

Sycamore fig
Ficus sycomorus
Sycamore figs can grow
to 20 m (66 ft) tall; this
specimen is found in
Ethiopia's Great Rift Valley.

It is the value of trees to humans that this book attempts to capture. Our human ancestors stopped swinging in trees and started walking on the ground between 4.2 and 3.5 million years ago, but the shift from arboreal to land-lumbering ape didn't mean leaving trees behind. Both our primitive and more recent ancestors will have climbed trees at night to escape the attentions of predators in much the same way that terrestrial primates like baboons do today; our early ancestors would also have climbed trees to forage for food, harvested bark for clothing and medicine, and used tree wood to make fire. For modern humans, trees continue to provide shelter and food together with a wide range of products of huge value to local livelihoods, national economies and global trade. Timber, fuelwood, wood pulp, medicinal and aromatic products, fruits and nuts are amongst the most valuable products derived from trees – many of which are sourced directly from the wild and are thus poorly recorded and undervalued.

This book goes far beyond the utility of trees: it is a true celebration of their existence. As Pope Francis said in his encyclical in 2015:

'It is not enough, however, to think of different species merely as potential "resources" to be exploited, while overlooking the fact that they have value in themselves. Each year sees the disappearance of thousands of plant and animal species which we will never know, which our children will never see, because they have been lost forever. The great majority become extinct for reasons related to human activity. Because of us, thousands of species will no longer give glory to God by their very existence, nor convey their message to us. We have no such right.'

Trees are a source of inspiration, of deep affection, spirituality and creativity. In the pages of this book, we celebrate the art and architecture that has been inspired by trees across nearly every human culture throughout history. From ancient Chinese artworks to children's fairytales to ultra-modern architecture, trees have been our muses, our protectors and our silent companions.

Seeds

Seeds

The biggest reproductive challenge for a tree is working out how to get the kids to leave home. Being literally rooted to the ground means that the mother tree must find inventive ways to transport her offspring as far away as possible, and to ensure that once they have found their own spot, they flourish.

01

Gyrocarpus americana illustration, 1795
Sometimes called the helicopter tree, *Gyrocarpus* fruits rotate, propelling the seed away from the mother tree.

Seed Dispersal and Structure

Trees disperse their seeds by packaging them so they can travel in both space and time. Adaptations such as wings and explosive pods enable short-distance dispersal; the winged fruits of the aptly named *Gyrocarpus americana*, for example, propel its seeds from the mother tree like mini-helicopters. Similarly, the pods of *Brachystegia* trees in central Africa's miombo woodlands twist as they dry until suddenly, with an explosive 'crack!', they shoot their seeds up to 30 metres (100 feet) – it's no coincidence that the local ChiBemba word for 'gun' is 'mfuti', the same name as the tree.

Longer-distance dispersal can be achieved by smaller, lighter seeds with long hairs that enable them to float in the wind, like those of the kapok tree (*Ceiba petandra*). To travel further still, seeds can hitch a ride on – or in – an animal. The large, spiny fruits of *Uncarina* in Madagascar are designed to adhere to fur and are so sticky they have been known to completely immobilize smaller animals. On one occasion, while I was collecting seeds in Madagascar, I came across the skeleton of a snake trapped in the harpoon-like grapple hooks of *Uncarina stellulifera*. Given the large size of *Uncarina* fruits, it has been suggested that they may have been dispersed by Madagascar's flightless elephant birds (*Aepyornis*) until their extinction about 1,500 years ago. More familiar to readers in the northern hemisphere will be the burr-like fruits of the sweet chestnut (*Castanea sativa*). Hitching a lift on the fur or feathers of an animal or bird is termed 'epizoochory', while being eaten, transported and being spat out or passed in an animal's droppings is called 'endozoochory'.

> To travel further still, seeds can hitch a ride on – or in – an animal. The large, spiny fruits of *Uncarina* in Madagascar are designed to adhere to fur and are so sticky they have been known to completely immobilize smaller animals.

Endozoochoric dispersal requires seeds to be packaged attractively, but in a way that won't entice specialized seed-eating animals that can damage the seeds irreparably and prevent germination. It is this packaging that we have to thank for many of our tastiest fruits [pg.38]. The plant family *Rosaceae*, for example, provides us with plums, apples, apricots, peaches, pears, cherries and other delights, all with the expectation that we won't eat the seeds, but will instead spit them out somewhere that they might flourish. Producing fleshy, sweet fruits takes a lot of energy for the mother tree; a less-costly strategy is to produce seed appendages, such as seed 'arils', which are high in fat or protein and are often brightly coloured to attract birds or mammals [pg.36]. Africa's pod mahogany (*Afzelia quanzensis*) uses this method, producing a bright scarlet aril that attracts birds with its bright colour and can be eaten without harming the seed inside. Probably the most striking seed aril is that of Madagascar's traveller's palm (*Ravenala madagascariensis*), which produces a bright, electric-blue aril designed to appeal to its main disperser, the ruffed lemur – reputedly capable of only seeing shades of blue and green.

Water is another means of long-distance dispersal, but is only useful to trees that live near a river or the sea. The long, pencil-shaped seeds of mangrove species like the red mangrove (*Rhizophora mangle*) are dropped into the water by the parent tree, where they float away until they come to a shallow area suitable for germination. Perhaps the most successful of these long-distance dispersers is the coconut (*Cocos nucifera,* pg.26), which can cross oceans and live for years immersed in seawater.

> Perhaps the most successful of these long-distance dispersers is the coconut (*Cocos nucifera*), which can cross oceans and live for years immersed in seawater.

Seed Germination

Travelling long distances is one thing, but finding the right spot to settle down is a different problem. Seeds come in deceptively simple packages – usually comprising just a seed coat ('testa'), a food source ('endosperm') and an embryo – and yet from tiny acorns mighty oaks grow. Seeds are the toughest biological objects on Earth, in some cases living for thousands of years, and they take some waking up. Like most babies, seeds need warmth, water and light to grow. Coconuts, washed up onto a tropical shoreline with lots of light and warmth, send down roots immediately – the coconut is well adapted for this, carrying enough food in its endosperm to sustain a long and vigorous taproot, which grows down into the soil as much as 90 cm (3 feet) in search of fresh water. Many tropical forest trees follow this same approach – they produce large seeds with high fat and carbohydrate content, which enables them to quickly establish themselves in a patch of light on the jungle floor. In warm, wet habitats like this, the best strategy is to get growing as soon as possible and stretch up to find the light.

02 – Red mangrove
Rhizophora mangle
Red mangrove propagules are long and pencil-shaped. Though they resemble seed pods, they are in fact embryonic roots.

03 – Ash
Fraxinus excelsior
Winged fruits containing a single seed, like those of the ash, are called 'samaras'.

For trees that grow in more arid or cold environments, rapid germination is a much riskier prospect. For this reason, their seeds are adapted to dry out and remain dormant until the weather conditions are right. This kind of dormancy is called 'physiological dormancy', whereby germination is triggered by temperature (termed 'stratification' or 'vernalization') or by specific chemicals. For trees living in temperate regions, the worst possible outcome is that their seeds germinate in mid-winter when temperatures are too low for them to survive, so seeds of these tree species employ stratification. A good example is the seed of the ash tree (*Fraxinus excelsior*), which needs to be kept warm (24°C/75°F) for at least thirty days, then chilled to 4°C/39°F for at least sixty days, before it will germinate. Seeds that stay dormant in the winter at sub-zero temperatures also need to dry out, because if their moisture content is too high, they will freeze, the ice crystals turning them to mush. The germination of some savannah species like the mungongo tree of the Kalahari Desert (*Schinziophyton rautanenii*), meanwhile, is stimulated by the chemicals in smoke, giving the seed the advantage of sprouting immediately after a wildfire when all the competition has been burnt to a crisp.

The most common form of seed dormancy is actually 'physical dormancy', a strategy in which the hard seed coat needs to be worn away before water can penetrate the seed and germination can take place. The seeds of Africa's famous umbrella thorn (*Acacia tortilis*), for example, have evolved to pass through the gut of a giraffe or elephant, where the seed coat is broken down just enough by the animal's digestive juices to enable germination in its dung. Gardeners mimic this process by nicking the seed coats of their sweet peas, enabling water to penetrate the seed and trigger germination.

02

03

Seed Behaviour

Seed biologists separate seeds into two main storage 'behaviour' categories, based on their sensitivity to drying out (desiccation). Seeds that can survive for long periods in a dry state are called 'orthodox', while seeds that die if they dry out are named 'recalcitrant'. The term recalcitrant – or 'difficult' – relates to the fact that these seeds cannot be stored in conventional seed banks in sub-zero temperatures [pg.40]. For these species, scientists must find a way to dry their reproductive tissue sufficiently for the seed to stand freezing without water crystallization. This usually involves cutting out the embryo of the seed, applying a chemical 'cryoprotectant', drying the tissue and freezing it rapidly in liquid nitrogen at -184°C/-299°F. When the embryo is subsequently woken up, it must be given a food source, the function normally provided by the endosperm...in other words, it's complicated! Fortunately for seed conservationists, most plants produce orthodox seeds, which can be safely dried down to 6–7 per cent moisture content, frozen at -20°C/-4°F and stored for decades without significant loss of viability.

> Kew Gardens has an incredible collection of historical artefacts and amongst these were olive seeds from Tutankhamen's tomb – known colloquially as 'Tut's nuts'.

How Long Can a Seed Live?

During my time working at Kew Gardens' Millennium Seed Bank in Sussex, we tested the viability of the seeds in our vault every ten years by taking a small sample of seeds from the collection (usually one hundred seeds) and trying to germinate them. We would then compare the result against the same test carried out when the seed was first banked. So, if one hundred seeds germinated before banking and only ninety germinated after ten years of storage, you knew that there was a one per cent loss in viability each year – meaning that all the seeds would be dead in a century's time. The oldest seeds in our collection were only about forty years old, and they were doing fine... But what about really old seeds?

Seeds can remain viable for decades if they are kept dry and cool. These cassia seed pods date from 1871–1920 and come from India.

Kew Gardens has an incredible collection of historical artefacts and amongst these were olive seeds from Tutankhamen's tomb – known colloquially as 'Tut's nuts'. When we X-rayed the seeds, it was immediately obvious that they had deteriorated to such an extent that they would never germinate. We also trawled through Kew's herbarium collections for seeds of pressed plant specimens that were hundreds of years old, but never succeeded in waking any of these up.

Our breakthrough came when we were contacted by a Dutch historian who was researching the life of an 18th-century silk merchant named Jan Teerlink. Teerlink was on his way back to Holland from the Dutch East Indies in 1803 aboard the SS *Henrietta*; during the voyage, he stopped off in Cape Town, where he collected forty small packets of seeds. At that time, Britain and Holland were at war, and on his way home Teerlink's ship was intercepted by a British frigate in the English Channel, resulting in the confiscation of all of his goods and a spell in the Tower of London for both him and his seeds. He was later released but his wallet and the seeds in it stayed there for over a hundred years before finding their way to the Admiralty and then on to the National Archives in Kew.

When the researcher discovered the wallet in the National Archives in 2006, he contacted us at the Millennium Seed Bank and asked if we thought any of the seeds would still be alive. Our first reaction was 'No chance': they had been stored for more than two hundred years in less-than-ideal conditions, and it was extremely unlikely that any of them would have survived. Nevertheless, we were willing to give it a go and obtained permission to try and germinate them. To our immense surprise, the seeds of three species germinated. Two of these were trees – a *Leucospermum* (conebush) and an *Acacia* species (later found to be *Acacia karoo*).

04

Kapok
Ceiba pentandra
Long-distance dispersal
can be achieved by smaller,
lighter seeds, with long hairs
that enable them to float in
the wind, like those of the
kapok tree.

We were able to grow the *Leucospermum*, and it is now a small tree in Kew's Temperate House. Better still, I had the opportunity to take ten cuttings of this tree back to Cape Town in 2013 to celebrate the centenary of Kirstenbosch Botanical Garden, which stands very close to the site from which the seeds were originally collected more than one hundred years before Kirstenbosch was established.

When we published the story of the two-hundred-year-old seeds, we received a letter from a lady who had worked in the herbarium of the Natural History Museum during the Blitz, where she recorded a similarly serendipitous discovery. In 1940, the Luftwaffe dropped an incendiary bomb on the museum, which set light to the herbarium. When the London Fire Brigade extinguished the fire, the water from their hoses germinated some lotus (*Nelumbo*) seeds on a preserved specimen that had been collected in China 170 years previously. In fact, our record for the world's oldest viable seed was short-lived. In 2005, Elaine Solowey, an Israeli biologist, managed to germinate some Judean date-palm (*Phoenix dactylifera*) seeds that had been found by archaeologists excavating Herod the Great's palace at Masada in 1963–65. These seeds were carbon dated by the University of Zurich and found to originate from 155 BCE to 64 CE, making them two thousand years old! The results of this study were published in 2008; Herod's seeds represent the oldest viable mature seeds recorded to date that have germinated naturally outside a laboratory. However, much older still was a cache of seeds found in ancient squirrel hibernation burrows 38 metres (125 feet) below the permafrost in north-eastern Siberia in 2011. These were carbon dated at thirty-two thousand years old, and scientists were able to extract the embryos of three *Silene stenophylla* (narrow-leafed campion) seeds and germinate them in the lab.

Coconuts

Water is a key form of long-distance dispersal, exhibited by plants like the coconut (*Cocos nucifera*). Adaptations for this purpose include being able to float, as well as the ability to survive in seawater for many months, protected by their outer husks.

← Coconuts and trees
Johannes Nieuhoff, 1618–72
Line engraving.

↓ Coconut
Cocos nucifera
Coconut trees provide food, fibre and building materials, and have formed the basis of island civilizations.

↘ Coconut kernel
The dried kernel of the coconut, from which oil is expressed, is called 'copra'. It is visible in diagram B.

↓ Dispersal
To stand a chance of establishing itself, the coconut must be swept above the tide line by a storm or spring tide.

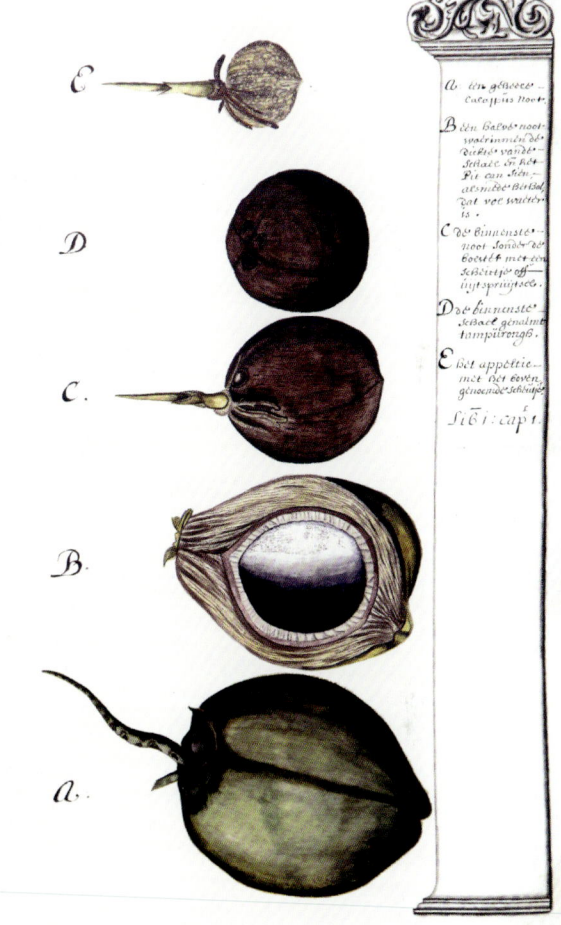

Architecture

Designed by local architecture firm Maison Edouard François, this apartment tower in Paris, France, has plants growing around its sixteen-storey exterior, creating a double skin. The studio notes, 'covered with plants from wild natural areas, our tower is a tool for seeding: it allows the wind to spread class-one purebred seeds into the urban environment'.

← **M6B2 Tower of Biodiversity, Paris**
Maison Edouard François, 2016
Stainless steel netting around the outer layer of the building provides a climbing frame for plants, whose seeds are spread across the city by the wind.

↓ The inner façade is made from recyclable green titanium panels, reminiscent of moss, that shimmer in the sun.

Size

Tree seeds vary enormously in size, from the diminutive dimensions of a pin head to over 30 cm (1 ft) in diameter. Larger seeds germinate very quickly, while smaller seeds can dry out and germinate years later. The former are less likely to be desiccation tolerant, and because they can't be dried, frozen and stored, such seeds are termed 'recalcitrant'. In contrast, smaller seeds can dry out, be stored at low temperatures and germinate years later. These seeds are referred to as 'orthodox'.

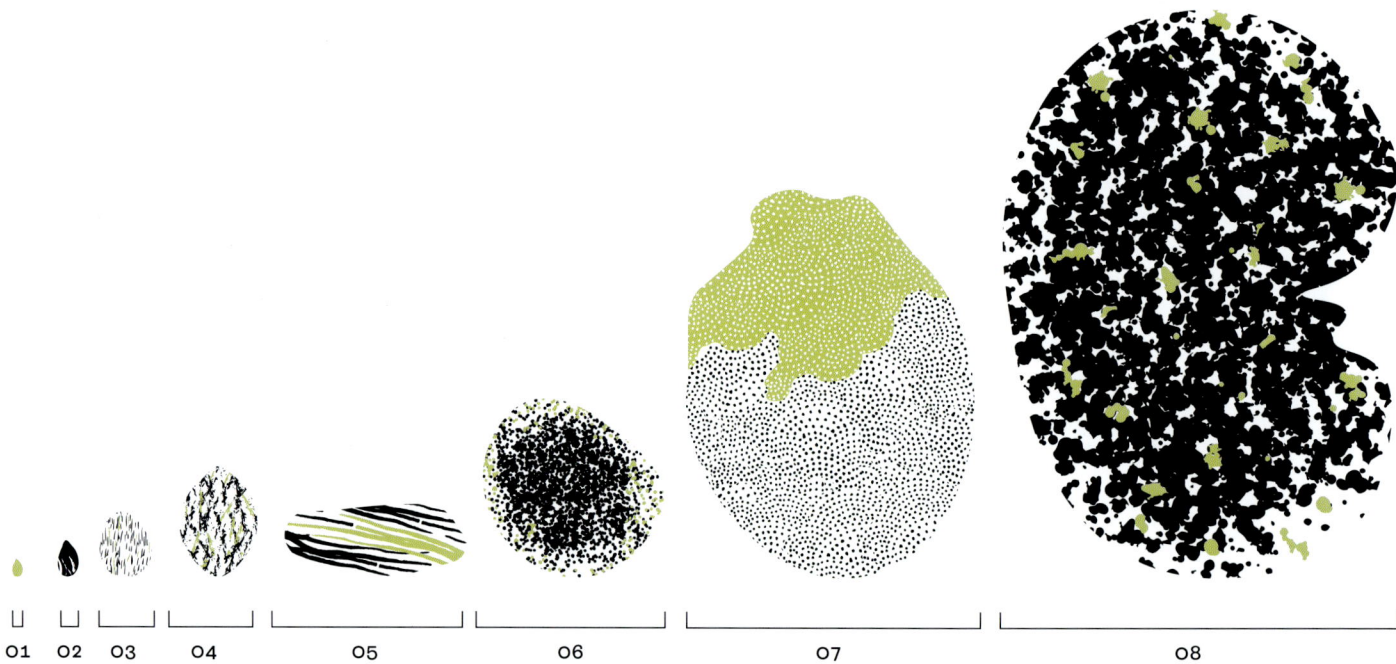

01   02   03   04   05   06   07   08

01 – Red spruce
Picea rubens
4 mm (0.16 in.)

02 – Apple
Malus domestica
8 mm (0.3 in.)

03 – Apricot
Prunus armeniaca
2 cm (0.8 in.)

04 – Peach
Prunus persica
3 cm (1.2 in.)

05 – Mango
Mangifera indica
7 cm (2.76 in.)

06 – California buckeye
Aesculus californica
7 cm (2.76 in.)

07 – African fan palm
Borassus aethiopum
11 cm (4.3 in.)

08 – Sea bean
Mora megistosperma
15 cm (6 in.)

09 – Coconut
Cocos nucifera
15 cm (6 in.)

10 – Double coconut
Lodoicea maldivica
30 cm (1 ft)

09

10

Dispersal Adaptations

Seeds are dispersed by the wind, water and animals. Their shape is usually a clue to their most common mode of dispersal.

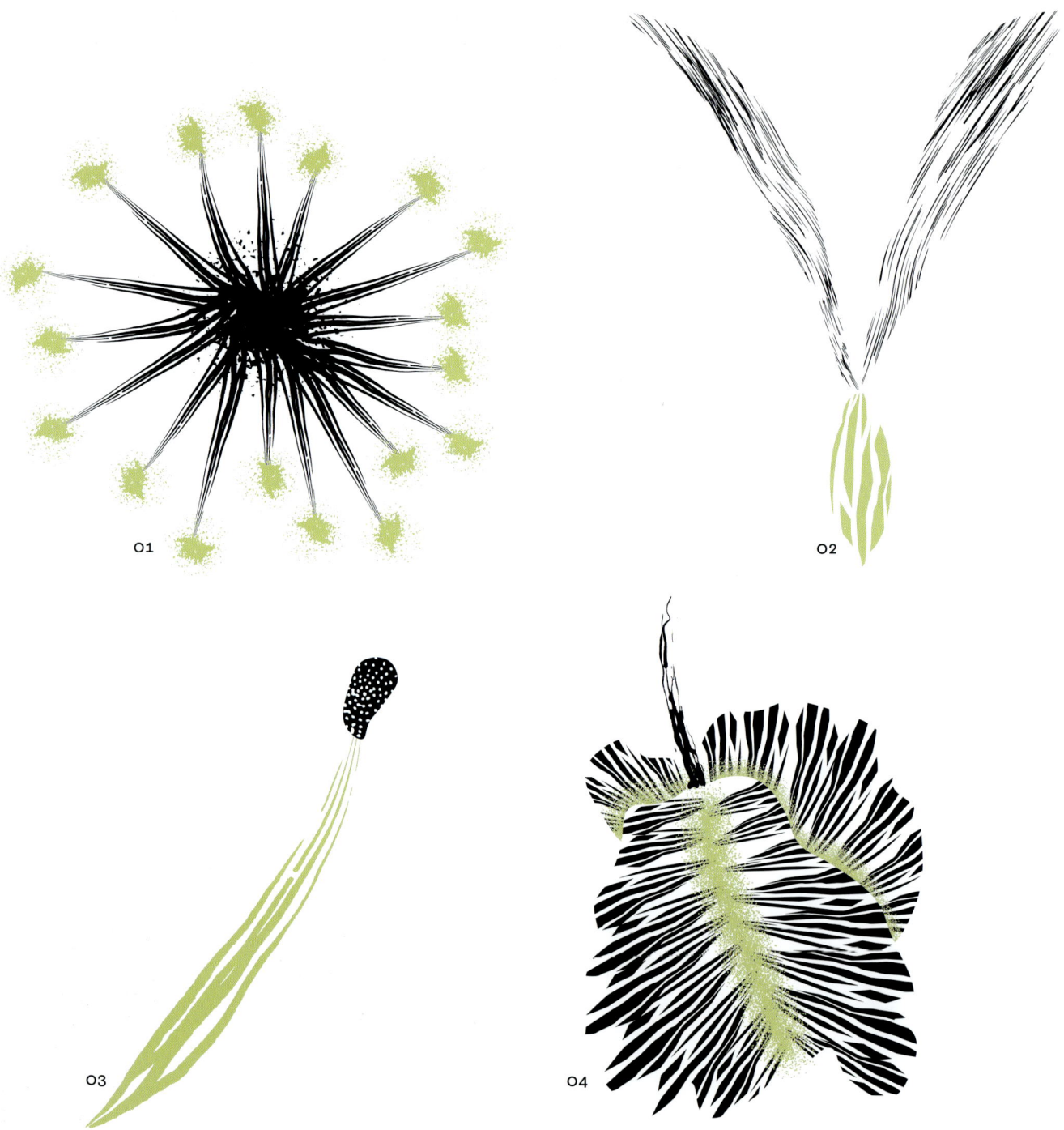

O1

O2

O3

O4

01 – Uncarina
Uncarina
The fruits of this Malagasy tree have hooked spines which catch in fur or feathers, enabling them to hitch a ride.

02 – Helicopter tree
Gyrocarpus americanus
The two-winged fruit of the helicopter tree spin like a propeller, enabling the seed to be dispersed considerable distances.

03 – Red mangrove
Rhizophora mangle
The spear-like fruit of the red mangrove either fall directly into the mud or float away to establish themselves somewhere else.

04 – Bushwillow
Combretum
Bushwillow fruits have four or five small wings that allow them to be swept along on the soil surface by gusts of wind for long distances.

05 – Clusterleaf
Terminalia sericea
Terminalia is closely related to *Combretum* but only has two wings, and doesn't catch the wind quite as efficiently.

06 – Sal
Shorea robusta
Shorea is a member of the *Dipterocarpaceae* family, famous for its hardwoods in Asia. 'Dipterocarp' means 'two-winged fruit'.

07 – Ash
Fraxinus excelsior
The single-winged fruit of ash spins and catches in the wind, helped by the fact that it is asymmetrical and weighted at one end by the seed.

08 – Sycamore
Acer pseudoplatanus
The sycamore is a member of the maple family, all of which have wind-dispersed, winged seeds.

Distance Travelled

Depending on their dispersal mechanism, seeds can travel a few metres to thousands of kilometres. This explains why some tree species have global distributions while others are restricted to small geographical areas.

01

02

03

01 – Helicopter tree
Gyrocarpus americanus
Propeller seeds – 10–20 m (33–66 ft). Heavy seeds with large wings rarely travel more than a few tens of metres.

02 – Zebrawood
Brachystegia spiciformis
Exploding pods – 30 m (98 ft). The pods of the zebrawood twist as they dry, eventually splitting with a loud bang and propelling the seeds a great distance.

03 – Kapok
Ceiba pentandra
Windblown seeds – 1 km (0.6 miles). The tiny seeds of the kapok tree are attached to long, feathery hairs that are blown long distances by the wind.

04 – Traveller's palm
Ravenala madagascariensis
Animal dispersed – 10km (6 miles). These bright blue seed arils attract lemurs, who eat them and disperse the seed in their faeces.

05 – Pod mahogany
Afzelia quanzensis
Bird dispersed – 100+ km (60+ miles). The red seed arils of the pod mahogany are carried away and eaten by birds, who leave the seed unharmed.

06 – Coconut palm
Cocos nucifera
Water dispersed – 1,000+ km (600+ miles). Coconuts can cross entire oceans before they are washed up and take root.

04

05

06

Ceiba pentandra, Ravenala madagascariensis, Afzelia quanzensis, Cocos nucifera

Colour

Different seed colours appeal to a variety of seed dispersers. Birds, in particular, are attracted to brightly coloured seeds – such as the lucky bean and rosary pea seeds shown below.

Lucky bean tree
Erythrina lysistemon

Bastard poon tree
Sterculia foetida

Areca palm
Areca catechu

Tea plant
Camellia sinensis

Ginkgo biloba
Ginkgo

Coral tree
Erythrina crista-galli

Black mangrove
Avicennia germinans

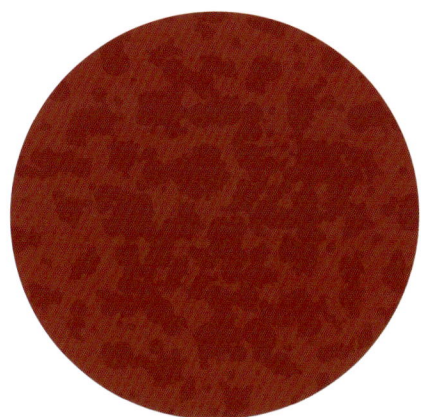

American chestnut
Castanea dentata

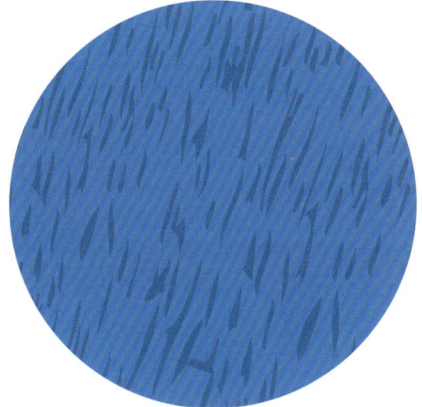

Traveller's palm
Ravenala madagascariensis

Sweet cherry
Prunus avium

Rosary pea
Abrus precatorius

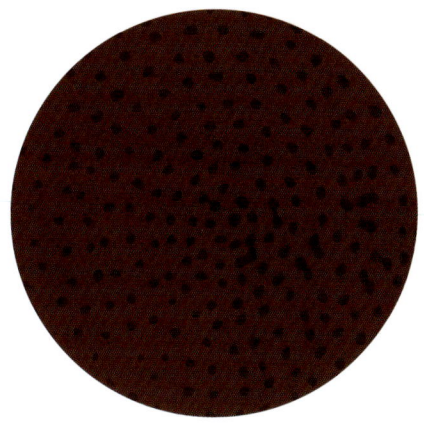

Japanese barberry
Berberis thunbergii

Fruits

Fruits are the receptacles of seeds, and so they also play an essential role in seed dispersal – the key difference being that they can be discarded once the seed has been dispersed.

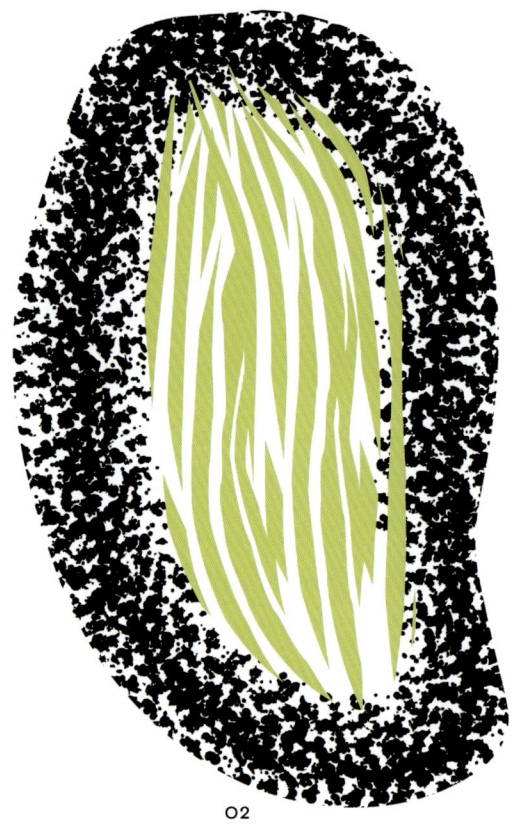

O2

O1

O3

O5

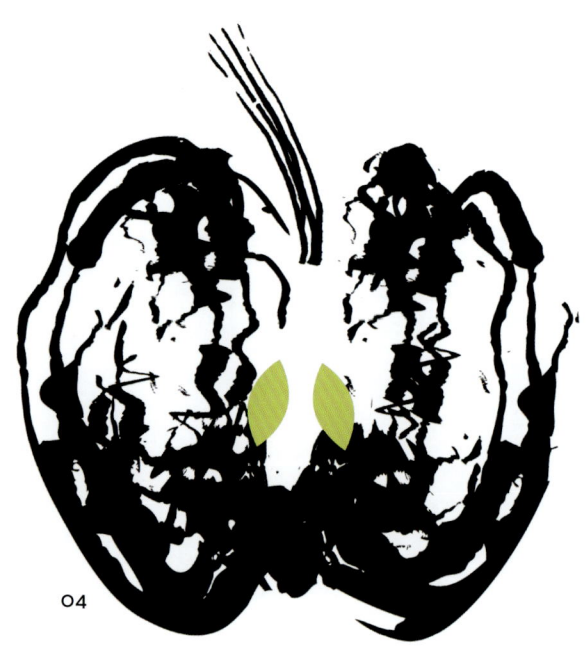

O4

38

01 – Orange
Citrus sinensis
Citrus fruits are segmented berries and are called 'hesperidia'.

02 – Mango
Mangifera indica
Mangos are 'drupes', which have a hard stone surrounded by soft flesh.

03 – Grape
Vitis
The grape is a 'berry', which has more than one seed embedded in a fleshy pulp.

04 – Apple
Malus domestica
The fruit of the apple is called a 'pome' – a fruit with relatively hard flesh that surrounds a core containing seeds.

05 – Plum
Prunus domestica
The plum is another drupe, with a hard stone enclosing the seed kernel.

06 – Apricot
Prunus armeniaca
Like the plum, the apricot is a drupe belonging to the rose family *Rosaceae*.

07 – Nectarine
Prunus persica var. nuci-persica
Nectarines are a kind of peach and, like all *Prunus* species, their fruits are drupes.

08 – Cherry
Prunus avium
Related to the peach, apricot and plum, cherry fruits are drupes.

09 – Blackcurrant
Ribes nigrum
Like the (unrelated) grape, the fruit of the blackcurrant is a berry.

10 – Pear
Pyrus communis
Like the apple, to which it is closely related, the pear fruit is a pome.

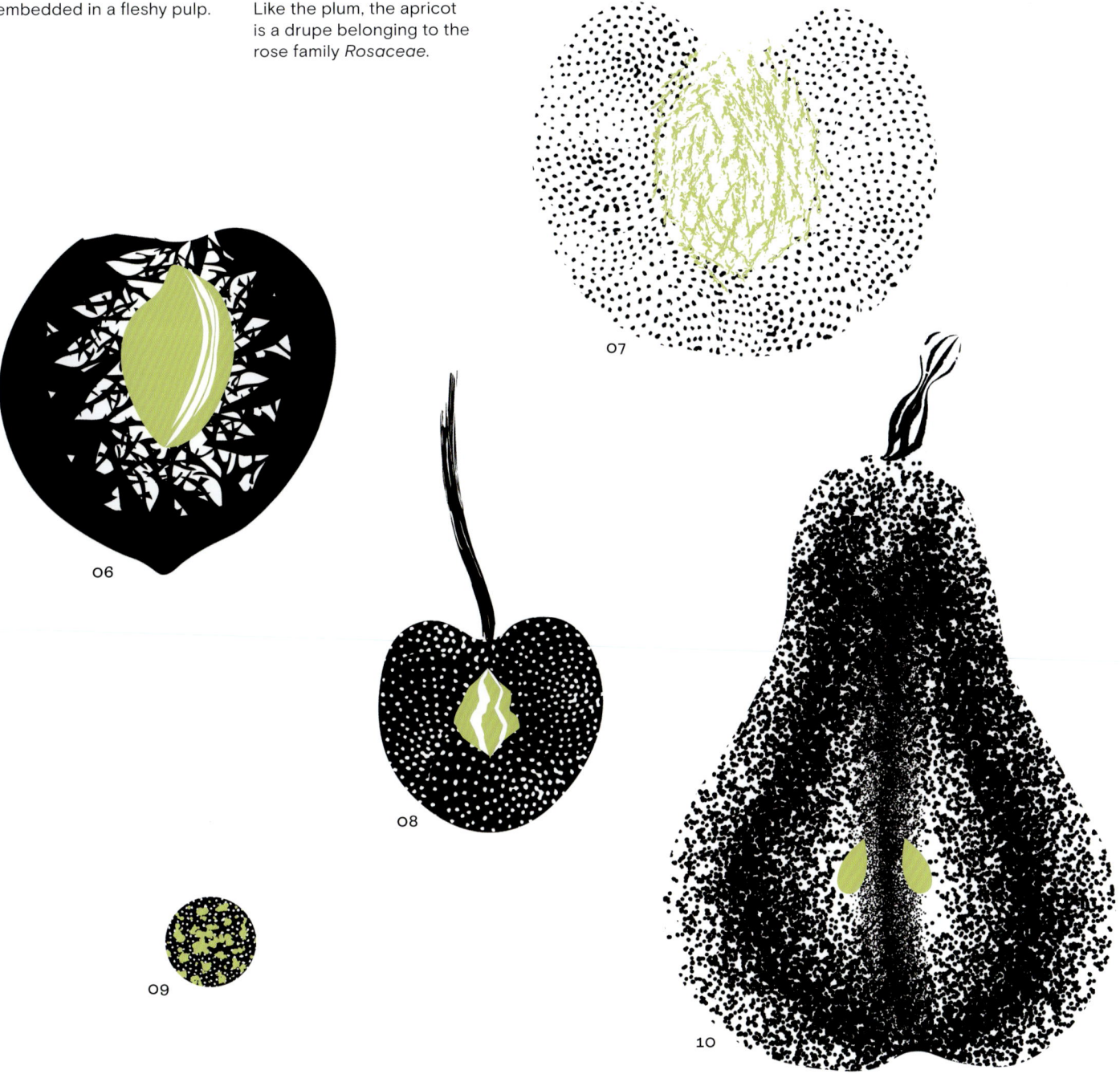

06

07

08

09

10

Seed Banks and the Seed Cathedral

The concept Heatherwick Studio masterminded was an audacious 'Seed Cathedral' in which multi-coloured seeds were embedded in the ends of sixty thousand acrylic rods.

People have been storing seeds for millennia, keeping them from one season to the next, either for their own use or for exchange with other farmers. In the early 20th century, crop seed banks (or 'genebanks') were developed as 'libraries' of seed collections for crop-breeding purposes. Using these seed collections, scientists can cross different cultivars and breed new crops, selecting them for specific traits such as disease resistance, increased yield or drought tolerance. The father of crop seed banks was Nikolai Vavilov, a Russian botanist whose life's work was identifying the historical centres of the origins of crops, and collecting seeds from across the globe for breeding purposes. His collections were amongst those stored in the Academy of Agricultural Sciences' seed bank in Leningrad which, from 1942–44, was besieged by German forces in the Second World War. Famously, the scientists in charge of the seed bank starved to death rather than eat the precious seeds inside it. Vavilov himself died in 1943 in one of Stalin's prisons, having fallen out of favour with the Soviet leader.

Since Vavilov's time, crop seed banks have proliferated, and crop breeding, together with fertilizers and pesticides, has formed the basis of the Green Revolution, enabling us to feed seven billion people worldwide. Today, more than 1,700 gene banks hold crop seeds for breeding purposes; the best known of these is probably the Svalbard Seed Vault, located on the remote island of Spitzbergen in the Arctic Circle. This unmanned facility was constructed by burrowing deep into a mountainside, taking advantage of the low temperatures of the permafrost to maintain the seeds at the low temperature needed for their long-term survival. Opened in 2008, the Svalbard Seed Vault currently holds more than a million duplicate seed samples from all over the world, providing a 'back-up' insurance policy in case those crops should be lost elsewhere.

The crops that we eat represent a tiny fraction of plant diversity. In fact, around 80 per cent of the calories and proteins we derive from plants come from just twelve species, while more than 50 per cent comes from wheat, rice and maize alone [pg.212]. In contrast, the total array of plant diversity equates to something like 400,000 plant species, of which about 60,000 species are trees. Many of these species – perhaps as much as a third – are directly useful to humans and, following the Rio Earth Summit of 1992, concern about the loss of biodiversity catalyzed a movement to conserve non-crop species in seed banks.

The most ambitious non-crop seed-banking initiative was the Millennium Seed Bank Project, established, as its name suggests, in 2000 by the Royal Botanic Gardens, Kew. Kew Gardens itself is located on a floodplain next to the River Thames in London so, instead of building the seed bank there, the Millennium Seed Bank (MSB) was constructed 100 metres (300 feet) above sea level at Wakehurst Place in Sussex, 100 km (60 miles) to the south of Kew. This £18-million facility comprises bomb-proof, flood-resistant underground vaults together with state-of-the-art laboratory facilities and a central atrium from where visitors can see the scientists at work. The design of the MSB mimics the rolling hills of the Sussex Weald in which it sits and, as well as being an iconic building, the Millennium Seed Bank is highly functional, storing more than 2.4 billion seeds from over 40,000 plant species. Working closely with affiliated institutions all over the world, the Millennium Seed Bank Partnership, as it is now called, has inspired a proliferation of 'wild' seed banks, including the architecturally inspiring PlantBank near Sydney in Australia – the largest seed bank in the Southern Hemisphere.

Svalbard Global Seed Vault
Spitzbergen, Arctic Circle
The Svalbard Seed Vault is
an un-manned crop seed
bank kept permanently cold
by the permafrost.

In 2009, as Head of the Millennium Seed Bank, I hosted a visit from the British designer Thomas Heatherwick, who was looking for inspiration for a commission he had won to design and build the UK Pavilion at the Shanghai Expo in 2010. The concept Heatherwick Studio masterminded was an audacious 'Seed Cathedral' in which multi-coloured seeds were embedded in the ends of sixty thousand acrylic rods, visible to visitors in the cathedral's interior and extending outside the building like quills on a porcupine's back. The challenge for us at the seed bank was to source the hundreds of thousands of seeds he needed. Seeds were sent by our partners from all over the world to achieve the task, only to find them refused entry by Chinese Customs upon shipping to Shanghai. The MSB's seed morphologist, Wolfgang Stuppy, was dispatched urgently to China to work with our friends and colleagues at the Kunming Genebank of Wild Species in southwest China to source all the seeds from within China instead. This was accomplished in a few short weeks, and the building became a reality. The Pavilion was as much an art installation as an exhibition centre and, with minimal interpretation, visitors dubbed it the 'Seed Cathedral'. The Pavilion was visited by 8 million people during its six-month life, and it won a RIBA International Award, the RIBA Lubetkin Prize and the London Design Medal.

Millennium Seed Bank

Kew's Millennium Seed Bank at Wakehurst Place in Sussex is the largest, most diverse seed bank in the world, with seeds from more than 40,000 plant species stored there as an insurance policy against extinction and as a source of seed material to support research.

← Seed storage
Seeds in the Millennium Seed Bank are stored in sealed glass jars at -20°C (-68°F), under which conditions they can live for hundreds of years.

↓ MSB buildings
Visitors can see scientists at work in the glass atrium, while the seeds are stored below ground in flood-, bomb- and radiation-proof vaults.

Heatherwick Studio's Seed Cathedral

The British Pavilion at Shanghai Expo 2010 was designed by Heatherwick Studio. The remarkable structure featured more than 200,000 seeds sourced from the Germplasm Bank of Wild Species at the Kunming Institute of Botany.

← From the outside, the acrylic rods extended from the building like quills on a porcupine's back. The grey exterior was transformed by coloured lights at night.

↘ **Seed Cathedral, Shanghai**
Heatherwick Studio, 2010
The interior of the Seed Cathedral comprised an array of 60,000 acrylic rods, each embedded with multi-coloured seeds and through which the light shone, creating an extraordinary effect.

Leaves

Leaves

Tree leaves are often a shade of green that we tune out or barely notice – the soothing, reassuring backdrop to the countryside or the nebulous blobs on brown sticks of a child's painting. Some are briefly glamorous in the autumn and then revert to type next spring. It's a pity we don't pay them more attention, because tree leaves are extraordinary. They feed us, provide us with oxygen, cure our illnesses, hide us and inspire us.

01

Some leaves are ephemeral, lasting no more than a few weeks, while others live longer than we do. Each is a food and oxygen factory that helps make our planet habitable. There are also an awful lot of them. A mature tree might have as many as 200,000 leaves; multiply that by the roughly 3 trillion trees on Earth, and that equates to around 85 million leaves per person.

As every schoolchild knows, leaves are where photosynthesis happens: where carbon dioxide and water are turned into sugars and oxygen, a process driven by the energy from the sun. This ability to manufacture food is called 'autotrophy' and it means that, despite their imposing statures, trees are at the base of many food chains. This makes them indispensable to a multitude of creatures. Scientists have discovered, for example, that around 2,300 species of invertebrates, mammals, birds, lichens and fungi rely to some degree on *Quercus robur*, the English oak, and this figure doesn't include bacteria and many other micro-organisms. Not all are directly part of the leaf-based food chain, but a significant number are, ranging from caterpillars that munch the leaves to 'saprophytic' fungi that break them down in the soil. Among the animals that feed on trees directly, leaf-eating mammals are termed browsers, though even the so-called grazers (eaters of grass and low-lying vegetation) will eat tree leaves now and then. Elephant and giraffe can reach up higher than other animals – into the tops of the tallest trees – meaning they don't need to compete with the multitudes of antelope that eat smaller shrubs. The giant birds of Madagascar, New Zealand and Australia were also leaf eaters, and trees have developed a wide range of mechanisms to keep such mega-herbivores at bay [pg.76].

Humans don't eat many tree leaves. While the leaves of the baobab or 'upside-down' tree of Africa and Madagascar (*Adansonia digitata*) for example are edible, and make a good spinach, they are hard to reach compared to the many kinds of terrestrial spinach on offer. They are also not available during the long dry season when food is most needed.

01 – Autumn Foliage
Tom Thomson, 1915
Oil on panel.

We may not eat many tree leaves, but we more than make up for this by drinking them. The global export value of tea was over US$6 billion in 2019; a third of those exports originated from China, where tea has been drunk for at least six thousand years. Like nearly all leaf-based infusions, tea was regarded by ancient cultures as having health benefits and was thus harnessed as a medicine. Many thousands of plant species are still used today in Ayurveda and traditional Chinese medicine and, although bark and roots are often regarded as more efficacious, leaves make up a significant component of this pharmacopeia. In my early years as a botanist in the Muchinga Mountains of Zambia, I recorded the use of various trees by traditional healers. One particularly peculiar remedy for epilepsy involved the healer chewing the leaves of the snake bean tree (*Swartzia madagascariensis*), and blowing into the mouth, nose and anus of the sufferer. Something that many traditional medicines have in common is that they are used as preventatives rather than cures. This approach mimics nature in that all animals, including carnivores, dose themselves with select plants to stay healthy – this is why dogs and cats sometimes eat grass and herbs.

As well as sustaining humans as a source of both food and medicine, tree leaves have provided humanity with inspiration for millennia. The leaves of the peepal fig (*Ficus religiosa*) are used as miniature canvasses for exquisite Indian art, while Indo-Persian art is rich in tree-leaf imagery and motifs, as evidenced by the paintings of Abul Hasan and the decorations on the Taj Mahal. Similarly, in China, Qiu Ying, one of the four Great Masters of the Ming Dynasty, painted exquisite trees, with individual leaves depicted in great detail. In Western art, the Arts and Crafts movement created a sense of leaf detail through light and colour. Perhaps the greatest exemplar of this was the comparatively little-known Canadian artist Tom Thomson, whose renditions of autumn leaf colours on Lake Algonquin perfectly capture the splendour of sugar maples and aspens [pg.66].

Leaf Deciduousness and Lifespan

'Deciduousness' – the dropping of leaves each year – is a tough choice for a tree because it involves a number of trade-offs. One of the key advantages is that the tree can pass the winter, when plummeting temperatures or low water availability make photosynthesis difficult, in a dormant state. With no leaves to catch the wind or hold the weight of snow, there is also less chance of breaking a bough and sustaining serious injury. Furthermore, when the spring arrives, producing flowers before the new leaves helps with wind pollination. On the other hand, ditching all the nutrients in its leaves in the autumn means the tree needs either to store those nutrients elsewhere or find additional nutrients in the soil to manufacture new leaves in the spring.

This nutrient trade-off also influences leaf lifespan. Generally, we think of leaves as being short-lived structures replenished annually or perhaps every few years, but this is not necessarily the case. The leaves of some evergreen tropical trees last for tens of years, as evidenced by experiments in forest canopies in which leaves are marked and found to still be there decades later. The main advantage of long-lived leaves is that precious nutrients are retained instead of discarded – this explains why evergreen species are dominant in infertile habitats where natural selection favours traits that preserve nutrients.

According to the *Guinness Book of Records*, the plant with the longest-lived leaves is the welwitschia (*Welwitschia mirabilis*). It is perhaps a stretch to call welwitschia a tree, but it does grow up to 1.5 metres (5 feet) tall. Found in the Namib desert, this extraordinary plant can live for up to two thousand years. It produces two leaves that grow continuously throughout its life, the largest leaves stretching more than 8 metres (26 feet),

wearing out at the tips and continually growing out of the basal plant. In a desert habitat, holding on to valuable nutrients is clearly an advantage, but the same is true in upland heathland with leached, acidic soils in which conifers are more competitive than their deciduous, broad-leafed counterparts.

Leaf Colour

In the case of deciduous trees, the wonderful shades of yellow, orange, red, purple and brown that we enjoy in the autumn months are very much part of the nutrient story too. The green colour predominant in leaves is the result of the pigment chlorophyll, the molecule that drives photosynthesis. Chlorophyll is associated with precious micronutrients such as magnesium and phosphates, as well as proteins that break down as the tree shuts down over the winter or the dry season. These nutrients are absorbed back into the tree stem and, as this happens, other pigments such as the yellow 'xanthophylls' and orange beta-carotene are revealed, giving the leaf its distinctive autumn colour. Another group of pigments, the 'anthocyanins', are produced towards the end of the summer as phosphates are reabsorbed by the tree, and these are responsible for the reds and purples that we see, particularly in species such as the sugar maple. The brightness of autumn colours depends on the species of tree and the weather, which influences the chemical reactions that take place in the leaf. Generally speaking, the brighter and cooler it is, the more spectacular the colours on show. Anthocyanins are also present in leaves that produce a spring flush, such as the miombo woodlands of south-central Africa, which cover an area of over 200 million hectares (495 million acres) and are more spectacular in the spring than the autumn.

> The leaves of some evergreen tropical trees last for tens of years, as evidenced by experiments in forest canopies in which leaves are marked and found to still be there decades later.

02 – Welwitschia
Welwitschia mirabilis
As listed in the *Guinness Book of Records*, the welwitschia has the longest-lived leaves in the world. Found in the Namib desert, this extraordinary plant lives for up to 2,000 years.

03 – Sugar maple
Acer saccharum
Anthocyanin pigments are produced towards the end of the summer, and these are responsible for the reds and purples that we see in the sugar maple.

Leaf Shape

Leaf shape [pg.68], texture and colour are all used by botanists to identify trees and other plants. Unlike flowers or fruits, leaves are present on a tree for most or all of the year, and are therefore good field characteristics for identification. There is a whole lexicon related to leaf shape, margins, texture, hairiness, colour and venation (the pattern of veins on a leaf), but given it's not particularly helpful to the non-specialist, this is where diagrams come in. Most humans are hard-wired to recognize leaf shapes and to distinguish one plant from another – presumably related to our hunter-gatherer history. Where it becomes a little tricky is in cases where plants are very closely related to each other and can only be differentiated by flower or fruit characteristics. Variability of leaf form in the same species or even on the same tree is another challenge for amateurs and specialists alike.

Given that learning the distinguishing characteristics of more than a handful of tree species takes some dedication and to become a real expert in a whole flora takes a lifetime, botanists have been exploring the bounds of automated identification. Leaf venation patterns can be useful to pick out species, and lend themselves to machine learning in which patterns are detected, learnt and assigned to particular plant species. Plant-leaf X-rays were used to investigate this possibility [pg.70], but of course X-raying your leaf in the field is not practicable. Instead, the very many plant-identification apps that are now available use a wide range of leaf characteristics gleaned from smartphone photos, including leaf shape, texture, colour and venation. In effect, every plant has a different spectral signature that is picked up by the camera and recognized by software through iterative machine learning. Typically, something like three hundred different images are needed to build up a signature for a particular plant species to be accurately identified.

03

02

Spectral Signatures

The spectral signature of a leaf is not just helpful to people with phone apps wanting to identify a particular plant in their garden or on a walk. It is also potentially valuable to ecologists and conservationists interested in the composition of forests. While satellite images and photographs from drones or aeroplanes are routinely used to estimate forest cover, extent and loss, they are not so useful when it comes assessing forest composition – an important indicator of forest health and conservation value. In recent years, scientists have been training satellites to recognize the spectral signatures of specific tree species. This requires matching the signature with verified, correctly named tree species, and botanic gardens have been invaluable in this respect. The world's botanic gardens and arboreta grow about 18,000 different tree species, which are accurately named and conveniently located for scientific study. However, with around 60,000 different tree species in the wild, we have a long way to go before we are able to characterize them all and automate their recognition from the air.

> Leaf hairs are another mechanism for conserving water; the denser they are, the more they reduce airflow around the stomata, thereby lowering transpiration rates.

04

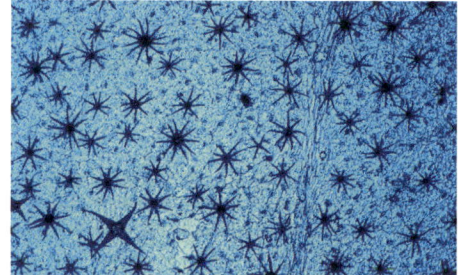

05

Leaf Texture

Leaf texture [pg.72] gives a good indication of the environment in which a tree grows. Waxy leaves with small surface areas, such as those found in tree cacti, tree *Euphorbias* and the extraordinary *Didiereaceae* family of Madagascar's dry forests and spiny thickets, are adapted to conserve water. The leaves of these succulent species also tend to have relatively few stomata – the mouth-like leaf structures that enable exchange of gases and water vapour. In addition, unlike plants in wetter environments, the stomata open during the cool desert nights in order to conserve water. Leaf hairs are another mechanism for conserving water; the denser they are, the more they reduce airflow around the stomata, thereby lowering transpiration rates. Other trees save water by modifying their leaves to minimize surface area and keep water loss down. These include conifers with needle-like leaves and some *Acacia* species that have developed 'phyllodes', which are actually modified leaf stalks. In both of these cases, the leaf surface to volume ratio is low, and water is conserved.

In contrast, trees that live in very wet environments can afford to transpire freely, losing water through multiple stomata that open during the warmer daylight hours. In very wet environments, leaf textures, shapes and arrangements are adapted to channel water away from the plant as much as possible. For example, many rainforest plants have drip tips and leaf veins and hairs arranged in such a way as to shed the maximum amount of water as quickly as possible. The leaves of some aquatic plants, such as the lotus (*Nelumbo*), are 'hydrophobic' and actually repel water. Here, dense, minute hairs covered with wax create a low-contact layer with the water droplets, preventing cohesion with the leaf surface and causing the droplets to roll off the leaf intact. Such structures have been recreated using nanotechnology to create 'self-cleaning' surfaces such as glass, coatings, paints, roof tiles and fabrics.

Leaf Defences

Leaves are the food-generating engines of trees. This means that leaves are full of nutrients – and that trees can't afford to lose too many of them. Given that trees spend their entire lives after germination in one place, they don't have too many options when it comes to avoiding animals that feed on their leaves. The term 'browser' – which we use to describe animals that eat tree leaves – might have rather benign connotations, but for trees, browsers are predators and robust defences must be employed to keep them at bay. The closest that trees come to escaping from browsers is growing so tall that most of their leaves are out of reach, particularly for larger herbivores. An elephant can eat up to 250 kg (600 lb) of fodder a day, and the black rhino can match this; the difference between the two is that the elephant can reach much, much higher – typically to a height of 6 metres (20 feet). Giraffes can exceed this height, so you clearly need to be quite a large tree to dodge leaf predators completely. Even very tall trees have to contend with arboreal browsers such as sloths, koalas and some primates, but these animals tend to eat much lower volumes of leaf material. An interesting specialized trait is the phenomenon of zig-zag branching found in some island ecosystems associated with giant birds. For example, in Madagascar, shrubs like *Decarya madagascariensis* produce small leaves arranged in a zig-zag pattern thought to be an adaptation to confuse and confound Madagascar's elephant birds – a now-extinct line of terrestrial birds that once weighed up to 750 kg (1,600 lb). We don't know for sure what elephant birds ate, but it is striking that similar zig-zag leaf arrangements are found in New Zealand, where the elephant bird's closest living relative, the kiwi, lives.

For smaller trees and shrubs where hiding leaves away or putting them out of reach is impossible, more aggressive measures are called for. Primary among these is the development of thorns, spines and prickles. Thorns are modified branches or stems, whereas spines are primarily modified leaf 'stipules' – small appendages that are usually found on a branch at the base of the leaf stalk.

Unlike stipules, though – which tend to be small, palatable, leaf-like structures – spines are hard, woody and sharp. Some spines have different beginnings, however; those found in cacti, for example, originate from leaf organs called 'glochidia'. Prickles, on the other hand, are more comparable to hairs in that they are derived from the cortex or epidermis and include spiny teeth on the margins of the leaves; an example of this is holly (*Ilex aquifolium*).

What all of these structures share is the function of deterring predators and, to this end, prickles, spines and thorns can be found on tree trunks, branches and leaves themselves. However, just as the giraffe has grown taller through natural selection to reach ever higher trees, it has also developed ways of dealing with the thorns and spines associated with its favourite food, the acacia tree. The giraffe's first defence is its thick, leathery tongue, impenetrable to thorns; moreover, its saliva has antiseptic properties, so that even if a thorn does cause injury, the cut won't become infected. This doesn't mean that predators have won the war. In fact, the more a tree is browsed, the greater the number of thorns and prickles it produces. Furthermore, physical defences are not the only weapon in the tree's armoury – it also engages in chemical warfare.

For smaller trees and shrubs where hiding leaves away or putting them out of reach is impossible, more aggressive measures are called for.

Many readers in the northern hemisphere will be familiar with the nettle (*Urtica dioica*) and its stinging hairs but, of course, nettles are not trees. So, think of a nettle scaled up both in size and fire power and you get *Dendrocnide moroides*, the 'gympie gympie' tree of Australia's east-coast forests, which grows up to 5 metres (15 feet) tall. As with all stinging plants in the nettle family, the blow is delivered by tiny, hollow silica hairs, which inject a toxin into the skin.

In the case of the gympie gympie, the toxin is a 'peptide' called 'moroidin', and the pain it induces has been likened to being burnt with hot acid and electrocuted at the same time. The most effective treatment is said to be rubbing the area with dilute hydrochloric acid to denature the peptide, and then using wax strips to remove the remaining hairs (if that's the cure, imagine how bad the pain must be!).

Despite these impressive chemical defences, a number of small marsupials, birds and insects eat the leaves of the gympie gympie tree, suggesting that all-out attack is not always the greatest strategy for a tree. More subtle chemical manipulation may be more effective, such as producing high concentrations of tannins and alkaloids, which not only make the leaf less palatable, but also interfere with an animal's digestion. This is a very common strategy and, generally speaking, the older a leaf is the more tannins and alkaloids it accumulates.

Perhaps the most interesting way trees use chemicals to deter browsers is via 'indirect defence', where a tree or plant provides a reward to another organism in return for keeping predators away. The best-known example of indirect defence is found in ant plants or 'myrmecophytes'. These plants attract ants by providing them with food in the form of extra floral nectaries and food bodies, which contain sugars, fat and protein. The bullhorn acacia (*Vachellia cornigera*) is an ant plant from Mexico and Central America that not only provides ants with food, but also with shelter in the form of hollow spines or 'domatia'. In return for food and shelter, the ants rush to the defence of the tree – releasing a 'pheromone' that causes all individuals in the vicinity to attack – if a threat is detected from insects, livestock or even humans. It is believed that other animals can detect the pheromone too, which encourages them to steer clear of the tree.

Although they are sedentary, trees are far from defenceless – worth bearing in mind next time you idly pick a leaf.

Biomimicry

The lotus is perhaps the best-known hydrophobic (water-repellent) plant on Earth. When rain falls, the waxy surface of its leaves means water droplets remain intact and roll off naturally, taking any contaminants along with them; this process of self-cleaning has been coined 'the lotus effect'.

← Sacred lotus
Nelumbo nucifera
It is advantageous for plants that live in very wet environments to be able to repel water or channel it away from the plant.

↓ Waterproof jacket
Humans have used nanotechnology to replicate the lotus's natural hydrophobic properties in textiles like waterproofs as well as glass, paint and surfaces.

Art

Some of the world's most famous artists and painters have made trees their subjects, capturing the beauty and movement of their leaves – from Vincent van Gogh to Georgia O'Keeffe, Gustav Klimt to David Hockney, Paul Gauguin to Wassily Kandinsky.

← Olive Trees
Vincent van Gogh, 1889
Oil on canvas. Van Gogh himself observed, 'What I've done is a rather harsh and coarse realism beside their abstractions, but it will nevertheless impart a rustic note, and will smell of the soil.'

↓ Olive tree
Olea europaea
Unusually for a broadleaf tree, the olive is an evergreen species, although unlike the tree itself, its leaves are not particularly long-lived.

Ginkgo
Ginkgo biloba
Ginkgo is often called a 'living fossil' because of its similarity to specimens that grew 270 million years ago.

Architecture

Tree leaves and venation patterns (the arrangement of veins in a leaf) are increasingly being used to inspire major architectural projects around the world. These designs often uphold principles of sustainability and include eco-friendly features such as solar panels and rainwater collection.

← Oasys+System,
Abu Dhabi
Mask Architects, 2020
The Oasys is a public space in the middle of Abu Dhabi, providing cool shade and respite from the city heat. Modules of palm tree-like structures have integrated functions, including nozzles on the underside of the palms that spray mist and solar panels on their roofs.

↓ Palm trees
Arecaceae
Palm trees, like these ones in Indonesia, have long been associated with tropical and exotic locations, and represent one of the earliest cultivated fruit trees. Today, they line the iconic Hollywood Boulevard and feature alongside many Modernist homes in the United States.

Tom Thomson and the Group of Seven

While most readers will be familiar with the paintings of gardens and nature rendered by Impressionists such as Monet, Manet and Sisley, and Post-Impressionists including Van Gogh, less well known are Canada's Post-Impressionists Tom Thomson and the Group of Seven. The Group of Seven comprised J. E. H. MacDonald, Frederick Varley, A. Y. Jackson, Lawren Harris, Frank Johnston, Arthur Lismer and Franklin Carmichael. Formed by MacDonald in 1920, most of the members had worked at graphic design firm Grip Ltd in Toronto before the First World War and, although not a member of the Group of Seven (he died in 1917), Tom Thomson had also been a Grip employee for a short while, knew most of the Group and was hugely influential. He travelled and painted with Jackson and Lismer in particular, and was supported and mentored in his painting career by MacDonald. MacDonald had been made Head of Design at Grip in 1907 after returning from a three-year stint in London, where he was exposed to the Arts and Crafts movement. Art historians ascribe different influences to Tom Thomson and the Group of Seven, including Post-Impressionism, Art Nouveau, the Arts and Crafts movement and Abstract Expressionism, but together they became the first major Canadian national art movement, producing paintings directly inspired by the Canadian landscape around them.

> Together they became the first major Canadian national art movement, producing paintings directly inspired by the Canadian landscape around them.

Most of Thomson's paintings were made in and around Ontario's Algonquin Park. Established in 1893, the park is a complex mosaic of lakes, rivers and woodland, combining recreational areas with logging concessions. Most striking in its spring and autumnal colours, Thomson spent months at a time canoeing and drawing the landscape, producing over four hundred oil 'sketches' on wooden panels, and completing around fifty canvases derived from them. Thomson's most famous paintings are of pine trees (*The West Wind* and *The Jack Pine*) but his depictions of the autumn colours of birch and aspens are probably best captured in paintings such as *Opulent October* and *In the Northland*. On the morning of the 8 July 1917, Thomson took his boat out on Canoe Lake, and that was the last time he was seen alive. His upturned canoe was found later that day, and his body was recovered eight days later. His watch had stopped at 12.14, and he had a 10 cm (4 in.) bruise on his temple. The coroner's verdict was death by drowning but rumours of suicide and even murder surfaced in later years. Most historians agree that there is no substance to these suggestions, and although he was known to be a competent canoeist and outdoorsman, his friends also spoke of his naivety when it came to nature. That naivety extended to his paintings, which stand out in their simplicity and stark beauty.

01 – Tamarack
Tom Thomson, 1915
Oil on wood. The tamarack, *Larix laricina*, is native to Canada and, like many larch species, its leaves turn to gold in the autumn.

02 – Spring Ice
1916
After an original oil painting by Tom Thomson, from *The Studio, Volume 114*.

03 – The Jack Pine
1917
After an original oil painting by Tom Thomson, from *The Studio, Volume 89*.

04 – March
c. 1916
After an original oil painting by Tom Thomson, from *The Studio, Volume 77*.

03

01

02

04

Shape

Botanists have a wide vocabulary to describe leaf shape in order to identify tree species. Some common examples are given here.

01 – African baobab
Adansonia digitata
Compound palmate/digitate (hand or finger-shaped).

02 – Bay laurel
Laurus nobilis
Simple, lanceolate (pointed at both ends).

03 – Snake bean
Swartzia madagascariensis
Compound, elliptic (oval-shaped, small or no point).

04 – Arabian coffee
Coffea arabica
Simple, obtuse (bluntly tipped).

05 – Silver birch
Betula pendula
Simple, deltoid (triangular).

06 – Bauhinia
Bauhinia
Bifoliate obcordate (heart-shaped).

07 – Olive
Olea europaea
Simple, oblong.

08 – Skunkbush
Rhus trilobata
Compound, trifoliate (leaflets in threes).

09 – Persian silk tree
Albizia julibrissin
Compound, bipinnate (leaflets in rows that further sub-divide into rows).

10 – Sacred fig
Ficus religiosa
Simple, cordate aristate (heart-shaped with a spine-like tip).

Coffea arabica, Betula pendula, Bauhinia, Olea europaea, Rhus trilobata, Albizia julibrissin, Ficus religiosa

Pattern

Water and minerals travel through tree leaves via a network of 'veins'; venation patterns are a useful characteristic for identifying a tree or plant. Plant-leaf X-rays show venation patterns with great clarity.

01

02

03

04

05

01 – Fig
Ficus carica

02 – Maple
Acer

03 – Ginkgo
Ginkgo biloba

04 – Oak
Quercus

05 – Hazel
Corylus avellana

06 – Mountain ash
Sorbus

07 – Red mulberry
Morus rubra

08 – Sweet chestnut
Castanea sativa

09 – Japanese maple
Acer palmatum

10 – Majesty palm
Ravenea rivularis

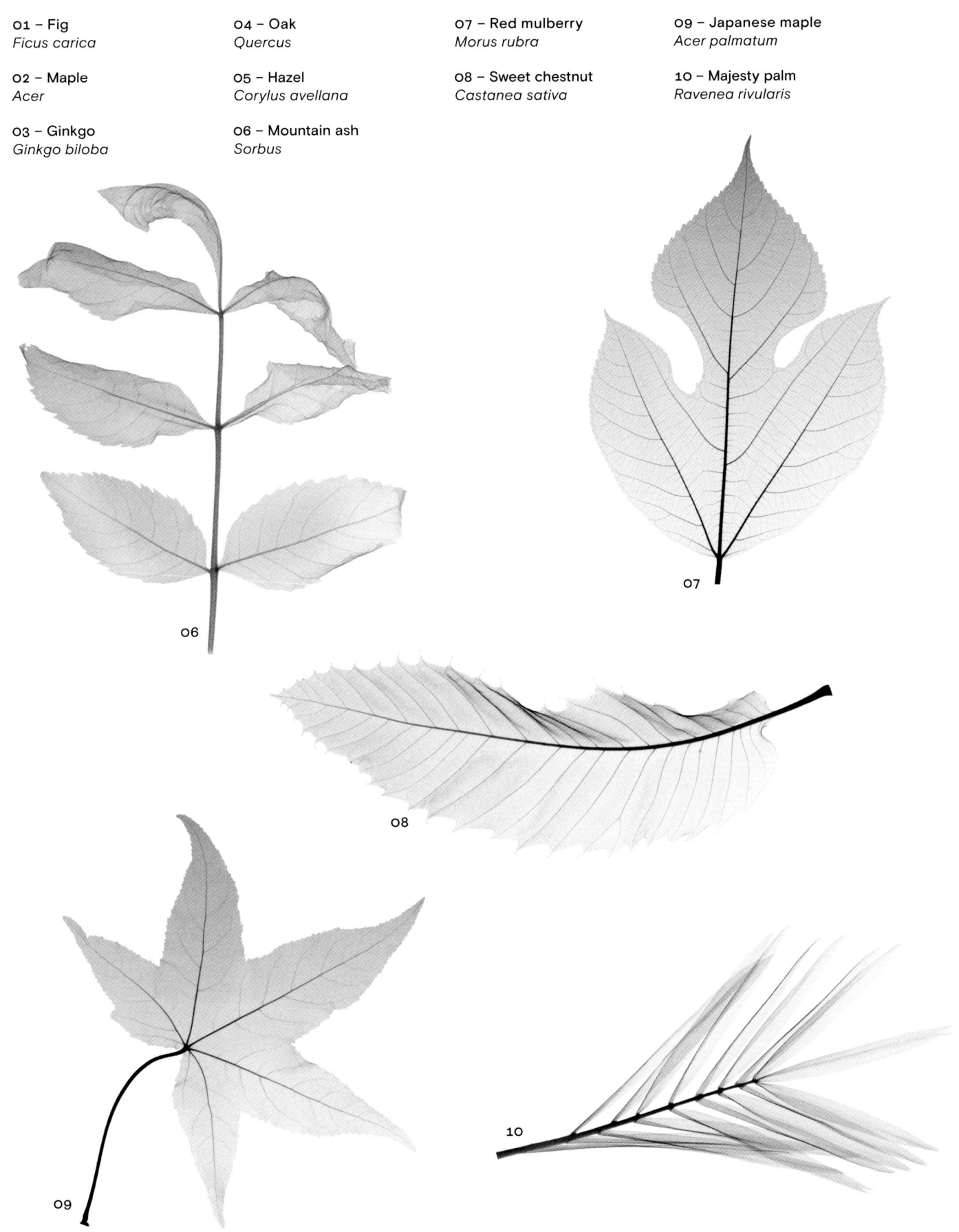

06

07

08

09

10

Texture

The range of leaf surfaces across species showcases more variety in adaptive traits. Small, waxy leaves conserve water, while in wetter climates, leaves are shaped to divert water away.

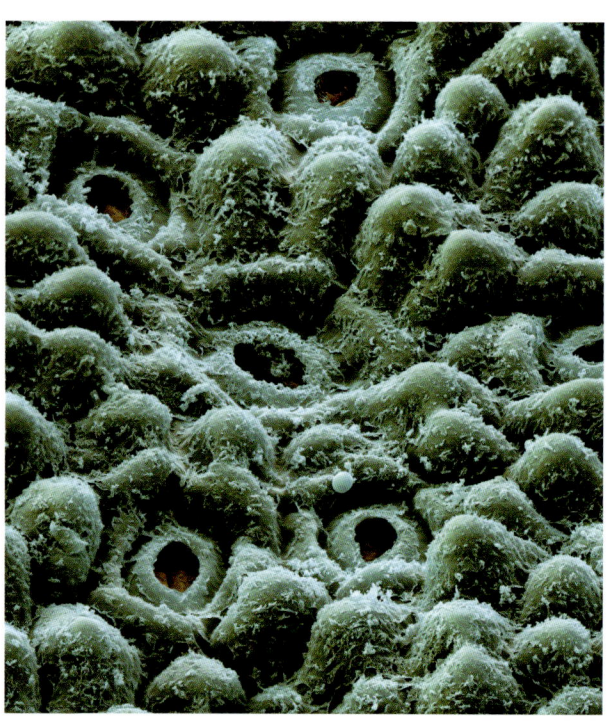

O1

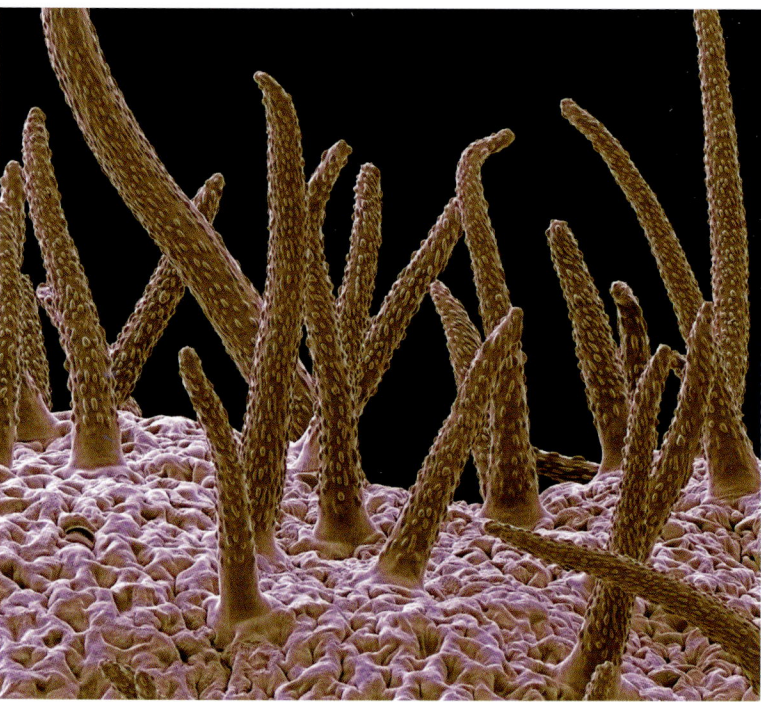

O2

O3

01 – Tasmanian
snow gum
Eucalyptus coccifera
Leaf surface micrograph
showing stomata.

02 – Horse chestnut
Aesculus hippocastanum
Micrograph of leaf surface
showing dense leaf hairs.

03 – Fuzzy deutzia
Deutzia scabra
'Stellate' (star-shaped) hairs
on the leaf surface.

04 – Olive leaf trichomes
Olea europaea
Leaf 'trichomes' – structures
that protect the leaf and
conserve water.

05 – Witch hazel
Hamamelis
Micrograph of witch hazel
leaf surface with stellate hair.

06 – Sweetbriar rose
Rosa rubiginosa
Globular hairs of the
sweetbriar rose.

04

05

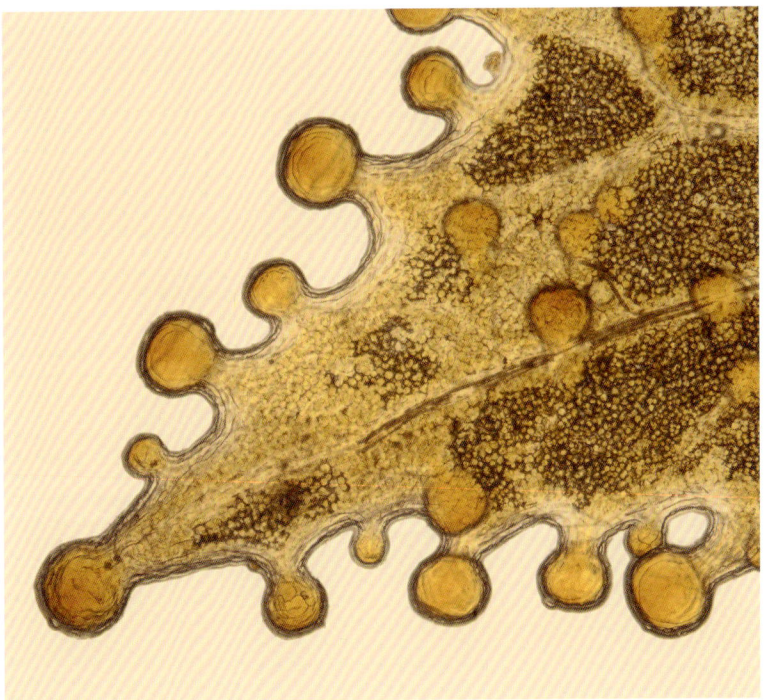

06

Deutzia scabra, Olea europaea, Hamamelis, Rosa rubiginosa

Defences

Many trees that grow low to the ground have developed formidable defences to repel leaf-eaters. Thorns, spines and prickles all serve this purpose, while thick, tough leaves can be equally off-putting to herbivores.

← **Blue thorn**
Acacia erubescens
This southern African tree has wicked, hooked thorns that mount an effective defence against smaller browsers but are less of a deterrent to larger species such as giraffe and elephant.

↓ **Madagascar ocotillo**
Alluaudia procera
Madagascar ocotillo is a succulent tree, native to south Madagascar. Its spines are arranged around the leaves as a defence against herbivores.

Form

Form

Trees take many forms, from towering redwoods to bulging baobabs, noble old oaks to spindly saplings. As with all things in nature, any definition of what constitutes a tree must take into account considerable variability, related to the environments in which different species grow and survive. Diversity in tree form doesn't stop at ground level either, but is replicated in the often complex root systems that tunnel beneath the surface.

01

01 – Yellow meranti
Shorea faguetiana
The tallest tropical tree,
up to 100.8 m (331 ft) tall.

The Definition of a Tree

The definition of a tree (as opposed to a shrub or other woody plant) used by the International Union for the Conservation of Nature (IUCN) is 'a woody plant with usually a single stem growing to a height of at least two metres, or if multi-stemmed, then at least one vertical stem five centimetres in diameter at breast height.' Even when a precise description like this is used, there is room for confusion and even controversy due to natural variation within species. For example, Africa's majestic lead-wood tree (*Combretum imberbe*) is usually seen as the epitome of a tree, growing to a height of over 25 metres (82 feet) with a single trunk and a rounded crown – akin to a child's 'lollipop' rendition of a tree. However, in certain habitats and soil types, the same species can grow as a multi-stemmed shrub, an 'ecotype' that bears little resemblance to the tree of the popular imagination. In fact, when early naturalists collected specimens of this ecotype, it was assumed to be a different species entirely, but has since been recognized as a variant. In order to take into account this kind of natural diversity in life form, IUCN's Global Tree Specialist Group includes all species that have been recorded *somewhere* as naturally growing trees in their global checklist of trees.

> In a rainforest where a tree is primarily competing for light, having a tall, straight trunk with a compact, rounded crown right at the top of the tree will be an asset.

A specimen called 'Thimmamma Marrimanu', which grows in Kadiri, India, has a canopy that covers an area of 1.9 hectares (4.7 acres), making it the world record holder for the breadth of its crown.

Tree-form Adaptation

The shape or form of a tree reflects certain characteristics that are useful adaptations in the particular habitats that trees occupy [pg.94]. For example, in a rainforest where a tree is primarily competing for light, having a tall, straight trunk with a compact, rounded crown right at the top of the tree will be an asset. Being this top-heavy will also require buttresses at the base of the trunk to help ensure stability, and root systems will be comparatively widely spread and shallow. Conversely, trees in more open, drier environments will be vulnerable to high winds and lightning strikes the taller they grow, and deep taproots will help them find water. For these species, shorter, diffuse, multi-stemmed shrub-like forms are better adaptive traits than a top-heavy trunk and crown.

Extreme weather such as high winds, heavy snowfall and floods represent the greatest threat to trees outside the activities of man. Adaptations such as the conical shape of boreal conifer trees (representing pines, firs and spruce) ensure that heavy snow slides off the trees, aided by the downward sloping angle of the branches and the waxy surfaces of the leaves. Regular flooding encountered by riverine or swamp tree species, meanwhile, has seen the evolution of special aerial roots called 'pneumatophores', through which gas exchange can occur. Various tree species have also developed surprising advantages to protect them against strong winds and gales. The Great Storm that swept southern England in October 1987 blew down an estimated 15 million trees, with wind speeds reaching 190 kph (120 mph).

The 'Sunland Baobab' was famous for housing a bar and wine cellar, and is one of many hollow baobabs that have been turned into dwelling places in Africa.

Amongst the survivors were the thousand-year-old oak trees in Windsor Great Park; curiously, the fact that these trees were hollow set them apart from the hundreds of thousands of younger oak trees that fell in the storm. In the early 1990s, my wife and I carried out some research in Zambia on mopane trees (*Colophospermum mopane*) that were hollowed out by saprophytic fungi, and we found that they were more resilient to elephant and storm damage than intact trees. Dan Janzen, a well-known tropical ecologist working in Costa Rica, had found the same thing with trees there, and in the early 1990s, when we were writing up our Zambian work, we visited Windsor Great Park and spoke to the park curator, Ted Green, who confirmed the same observation. These findings illustrate the benefit of hollow cylinders, which are lighter and stronger than solid trunks and therefore much less likely to be blown over. Symbiotic relationships are discussed more comprehensively in Chapter 8 [pg.238], but this is an example of an unexpected symbiosis in which a fungus benefits from extracting the nutrients in the heartwood of the tree, while the tree profits from the loss of above-ground weight.

02 – African baobab
Adansonia digitata
The African baobab is
the most widespread
of the iconic baobabs.

02

Tallest, Largest, Stoutest, Broadest, Oldest, Smallest

The tallest trees in the world are the coastal redwoods of California and Oregon (*Sequoia sempervirens*). 'Hyperion', the tallest tree in the world, is over 115 metres (380 feet) high and resides in the Redwood National and State Parks complex in California. The parks are also home to the second and third tallest trees in the world, appropriately named 'Helios' and 'Icarus'. The tallest tropical tree is a specimen of Borneo's yellow meranti (*Shorea faguetiana*), which measures a hefty 100.8 metres (331 feet), and is named 'Menara', followed closely by a specimen of Tasmania's mountain ash (*Eucalyptus regnans*) at 100.5 metres (330 feet), called 'Centurion'. For perspective, the height of any of these trees topping one hundred metres is equivalent to more than seven double-decker or Greyhound buses parked end to end, or a thirty-storey building.

The largest trees in the world, measured by the volume of their trunks, are the giant sequoias (*Sequoiadendron giganteum*), also found in the Pacific Northwest. The largest of these is 'General Sherman' at 1,487 cubic metres (52,500 cubic feet), and the top twelve largest trees are all giant sequoias. Next comes the coastal redwood 'Grogan's Fault' (1,084 cubic metres/38,281 cubic feet) and then a long way behind is a specimen of New Zealand's kauri (*Agathis australis*) named 'Tane Mahuta' at 516 cubic metres (18,222 cubic feet). Largest doesn't mean stoutest, and here the record goes to Mexico's Montezuma cypress (*Taxodium mucronatum*) – specifically a specimen called 'Arbol de Tule', a tree growing in the grounds of the church at Santa Maria de Tule in Oaxaca. This specimen measures 11.62 metres (38.1 feet) in diameter, provided its buttresses are taken into account. Without the buttresses, it measures 9.38 metres (30.8 feet), relegating it to second place, behind South Africa's 'Sunland Baobab' (*Adansonia digitata*) which, before a section of the tree died in 2017, measured 10.64 metres (34.9 feet) in diameter.

The 'Sunland Baobab' was famous for housing a bar and wine cellar, and is one of many hollow baobabs that have been turned into dwelling places in Africa. For example, Namibia's 'Ombalantu Baobab' can accommodate thirty-five people, and has been used variously as a chapel, house, hiding place and post office in its eight-hundred-year history. Measuring the diameter of a dryland tree like the baobab is not straightforward because baobabs expand in the rainy season as they take up water, then contract in the dry season. Baobabs also have a tendency to form multiple trunks, making it difficult to know whether you are measuring a single tree or multiple trees. An example of this is the 'Big Tree' near Victoria Falls in Zimbabwe, which is thought to be the largest diameter tree in Zimbabwe but may in fact be three separate trees that have grown together.

03

When explorer and missionary David Livingstone arrived at the Victoria Falls in 1855, he was convinced that the huge baobab trees he saw there were of enormous antiquity. As a follower of Bishop Ussher, Livingstone was perplexed, since the Bishop insisted that the

Unlike the world's largest trees, the smallest trees owe their existence entirely to the efforts of the people who cultivate them.

world had been created on 23 October 4004 BCE. In fact, baobabs are comparatively fast-growing considering their size, and carbon-dating techniques suggest that the largest specimens are little more than one thousand years old. When it comes to age [pg.98], baobabs are mere striplings compared to North America's bristlecone pine

03 – Bristlecone pine
Pinus longaeva
The oldest-known specimen of bristlecone pine – 'Methuselah' – is more than 4,850 years old.

04 – Quaking aspen
Populus tremuloides
A clonal colony in Utah is the heaviest organism in the world.

(*Pinus longaeva*). The oldest-known specimen of bristlecone pine – aptly named 'Methuselah' – is 4,853 years old, and lives in the White Mountains of California. This tree germinated from seed in 2833 BCE, a date that does contradict Bishop Ussher's chronology, as it predates the Bishop's calculation of when the Great Flood occurred (2349 BCE). The bristlecone pine grows as an individual tree rather than a clonal colony of trees.

Clonal colonies reproduce asexually from root suckers rather than from seed, with older parts of the colony dying off and being replaced with newer shoots. Genetically, these are the same organism, but they are constantly being renewed, and therefore can live for much longer than an individual tree. A clonal colony of quaking aspen (*Populus tremuloides*) in Fishlake National Forest in Utah is known as 'Pando' (Latin for 'I spread'), covers an area of 43.6 hectares (108 acres) and is estimated to weigh some 6,000 metric tonnes (5,900 imperial tons) – making it the heaviest known organism in the world. An age of 80,000 years has been postulated for Pando, but this is now thought to be unlikely given the colony would have had to survive the last ice age 'just' 10,000 years ago. Clonal trees that have been radiocarbon dated include 'King Clone', a creosote bush (*Larrea tridentata*) in the Mojave Desert (11,700 years old) and 'Old Tjikko', a Norway spruce (*Picea abies*) in Norway at 9,550 years old.

The oldest tree planted by people is thought to be the 'Jaya Sri Maha Bodhi' tree (*Ficus religiosa*), which grows in the Mahamewna Gardens, Anuradhapura, Sri Lanka, and was planted in 288 BCE. This tree was itself derived from a cutting of the tree under which Buddha attained enlightenment at Bodh Gaya in India in the 5th century BCE. Although *Ficus religiosa* grows to an impressive age and size, it is dwarfed by its close relative, the Banyan (*Ficus benghalensis*). This tree – specifically a specimen called 'Thimmamma Marrimanu', which grows in Kadiri, India – has a canopy that covers an area of 1.9 hectares (4.7 acres), making it the world record holder for the breadth of its crown.

Unlike the world's largest trees, the smallest trees owe their existence entirely to the efforts of the people who cultivate them. The term 'bonsai' is now widely used to describe the cultivation of miniature trees using techniques of root and crown pruning [pg.100]. Derived from the Chinese tradition 'penzai', bonsai (meaning 'tray planting') has become a Japanese art form widely practised by horticulturists around the world. Miniature bonsai trees are only a few centimetres tall, despite being mature trees, some of them hundreds of years old, and bearing fruits or flowers.

04

Dipterocarp
Dipterocarpus
These canopy trees are
amongst the tallest tree
species in Asia.

Below-ground Structures

The trunk and crown of a tree constitute its above-ground biomass, but below the surface its root system can make up a sizeable 20–30 per cent of the tree's total mass [pg.102]. Root patterns vary enormously depending on the species of tree, climatic conditions and soil type, but all roots fulfil the same basic function – that of providing micronutrients and water for the tree. They are aided in this task by symbiotic, 'mycorrhizal' fungi [pg.243], whose microscopic, hair-like 'filaments' called 'mycelia' spread into the tiniest soil particles, sequestering minerals and water from the soil in exchange for sugars and carbohydrates manufactured by the tree. Over the past few decades, scientists have discovered that these mycelia and root networks not only enable trees to exchange nutrients with fungi, but also with each other – on the face of it an altruistic action that seems hard to reconcile with the concept of the survival of the fittest. Studies have also shown that if one tree is attacked and injured, then the defences of other trees in the forest – connected to the first through mycelia – are similarly triggered; when the mycelial connections between these trees are broken, the other trees do not react. Most interestingly of all, it seems that 'mother' trees are able to recognize their own offspring and divert resources to nurture them.

> Root patterns vary enormously depending on the species of tree, climatic conditions and soil type.

Architecture

Trees are integral to architecture and design.
They are the protagonists of landscape
architecture and provide timber for building, but
tree form also informs construction methods and
interior design, from heavyweight religious and
cultural landmarks to quirky treehouses.

← Kapok tree
Ceiba pentandra
Many tall forest trees, like these specimens at Angkor Wat in Cambodia, have buttressed trunks that add to their stability, particularly if they are shallow rooted.

↙ Sagrada Família
Antoni Gaudí, 1882–
Gaudí used the same solid buttresses to support the inclined columns on the narthex of the Passion Façade of the Sagrada Família.

↓ Agri Chapel
Yu Momoeda Architects, 2016
The light and airy Agri Chapel, just outside Nagasaki, Japan, is surrounded by woodland. Its thoughtful design – using traditional Japanese construction techniques, with a pendentive dome supported by a series of tree-like columns – mimics its natural setting.

Landscape Architecture

In addition to being sources of raw materials for building, trees are also a source of inspiration for architectural designs. They have the advantage of combining beauty of form with strength and functionality.

Gardens by the Bay, Singapore
Grant Associates and Wilkinson Eyre Architects, 2012
In this garden, the iconic architectural 'trees' are cooling towers for the garden's biomass boilers.

There is an increasing awareness of the positive association between mental and physical health and access to green spaces. With 68 per cent of the world's population projected to live in urban areas by 2050, prioritizing people's access to trees and nature is of the utmost importance.

Climatic Adaptations

Trees are indelibly shaped by the environment
in which they grow. From elephantine trunks that
store water to tall, slender buttresses reaching
for light, variations in form help trees to survive.

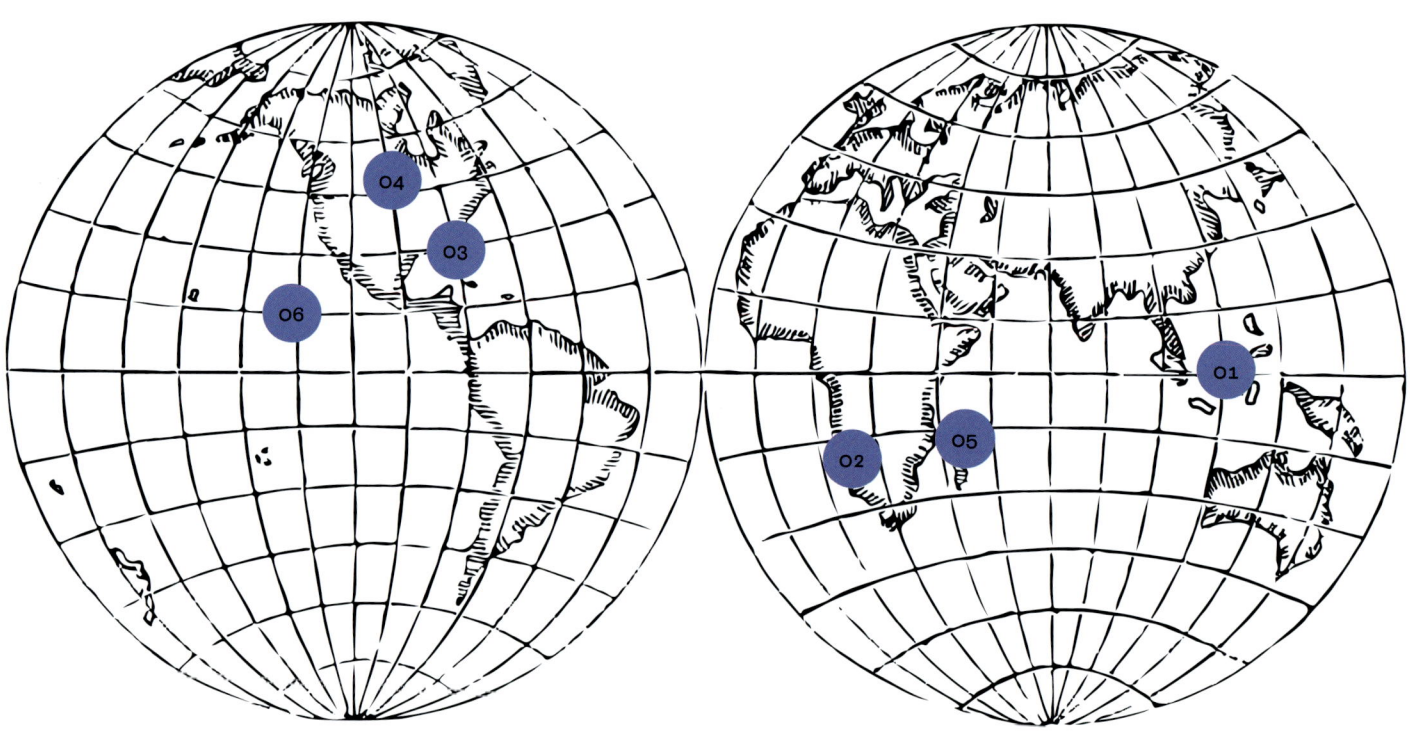

01 – Dipterocarp (*Dipterocarpaceae*), Malaysia – *Light*
A tall, straight trunk is an advantage when competing
with other species for light.

02 – Acacia (*Acacia*), Namibia – *Drought*
Desert trees usually grow in seasonal watercourses,
where water collects and persists below the surface.

03 – Red mangrove (*Rhizophora mangle*), USA – *Swamp*
Some waterlogged trees have special aerial roots called
'pneumatophores' through which gas exchange can occur.

04 – Fir (*Abies*), Canada – *Snow*
Snow slides off the angled branches of fir trees,
reducing damage from the weight of it.

05 – Baobab (*Adansonia*), Madagascar – *Water*
The spongy wood of the baobab tree stores water
during the dry season.

06 – Fig (*Ficus*), USA – *Wind*
Thick buttress roots help to anchor tall, top-heavy
forest trees, reducing instability in high winds.

Tree Crown Mapping

Garden designs depict trees from above to
indicate rough canopy size, density, size of leaf
and whether a tree is deciduous or evergreen.

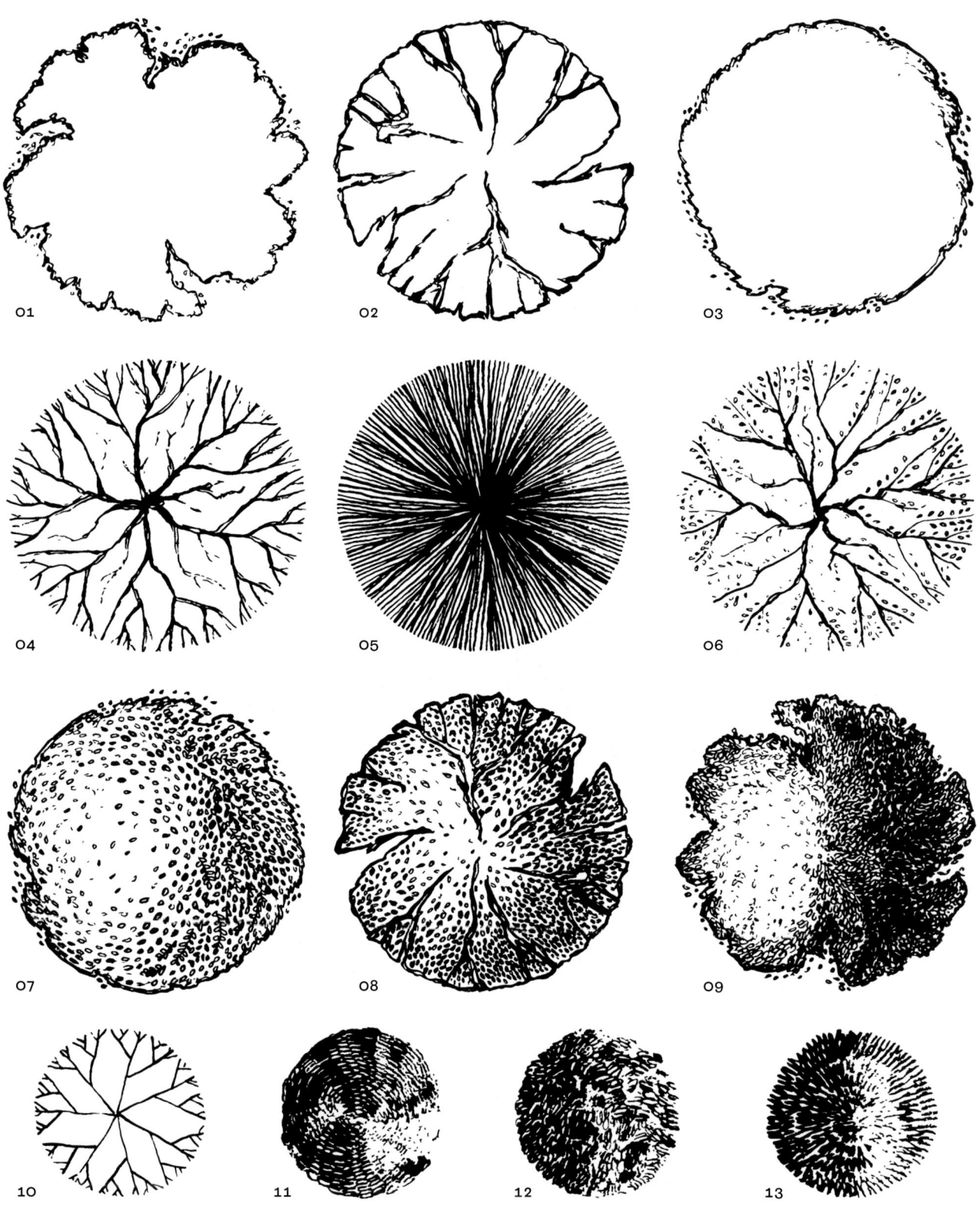

01 – Evergreen light
canopy profile

02 – Deciduous light
canopy profile

03 – Evergreen dense
canopy profile

04 – Deciduous
winter profile

05 – Evergreen
winter profile

06 – Deciduous
autumn profile

07 – Evergreen dense
canopy detail

08 – Deciduous light
canopy detail

09 – Evergreen light
canopy detail

10–13 – Smaller shrubs
Different textures, with
deciduous on the left and
evergreen on the right.

14 – As 04, Deciduous
winter profile
Tree-crown mapping is
used in garden design,
ecological descriptions
and profiling. In landscape
plans, rather than being
an exact rendition of a
particular species, these
pictorials give an idea
of what will grow in the
soil beneath.

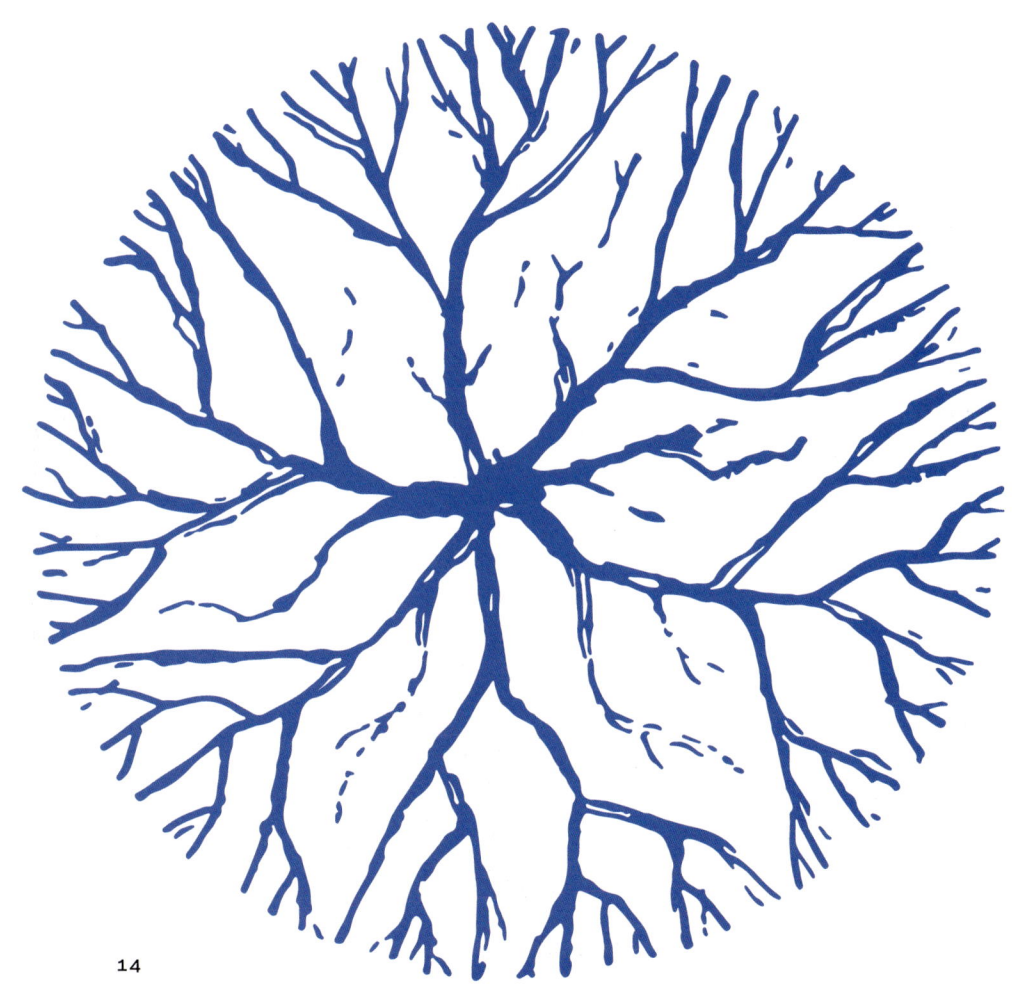

14

Oldest Living Species

A timeline of notable trees, organized by the
oldest-living member of each species.

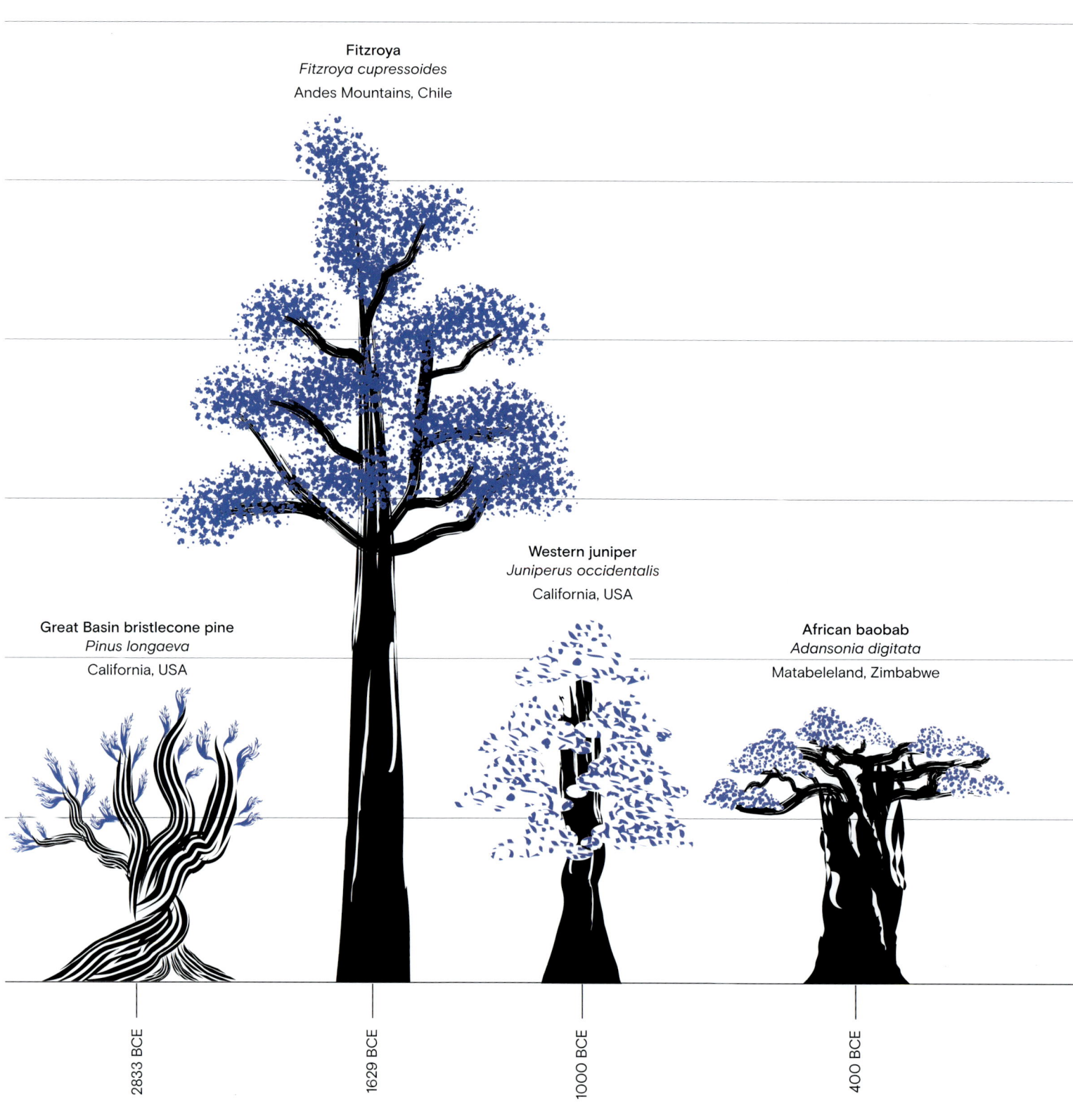

Fitzroya
Fitzroya cupressoides
Andes Mountains, Chile

Western juniper
Juniperus occidentalis
California, USA

Great Basin bristlecone pine
Pinus longaeva
California, USA

African baobab
Adansonia digitata
Matabeleland, Zimbabwe

2833 BCE

1629 BCE

1000 BCE

400 BCE

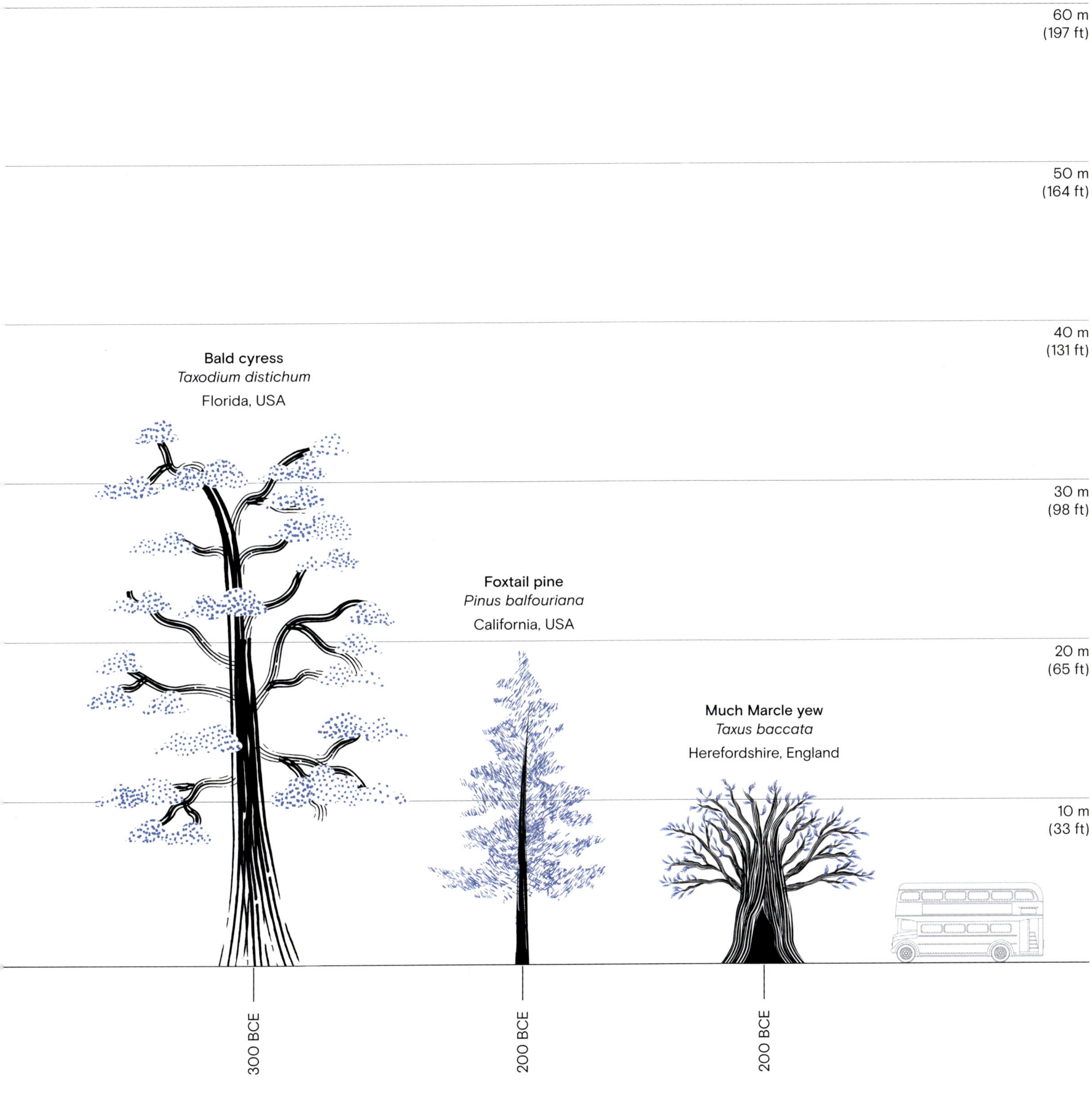
60 m
(197 ft)
50 m
(164 ft)
40 m
(131 ft)
30 m
(98 ft)
20 m
(65 ft)
10 m
(33 ft)
Bald cyress
Taxodium distichum
Florida, USA
Foxtail pine
Pinus balfouriana
California, USA
Much Marcle yew
Taxus baccata
Herefordshire, England
300 BCE
200 BCE
200 BCE

occidentalis, Adansonia digitata, Taxodium distichum, Pinus balfouriana, Taxus baccata

Bonsai Trees

'Bonsai' is Japanese for 'tray planting'; bonsais are miniature trees that mimic the shape and form of full-sized trees. Careful pruning determines the bonsai's final form.

Forest
Yose-ue
Comprised of multiple trees of various heights, rather than a single tree with more than one trunk.

Slanting
Shakkan
The trunk is angled to the right or left, with the branches trained to balance the weight.

Formal upright
Chokkan
The main trunk is straight and pointed upwards, as well as being wider at the base and tapered towards the top.

Broom
Hokidachi
A straight, upright trunk gives way to fine branches in all directions.

Literati or bunjin
Bunjingi
An approach rather than a set of strict criteria; slender trees with an abstract, calligraphic quality.

Root-over-rock
Seki-joju
Bonsai roots are grown so that they grip a rock or a stone.

Twin trunk
Sokan
Two upright trunks are
grown from the same root,
usually with one taller and
thicker than the other.

Informal upright
Moyog
The trunk is shaped like
a letter 'S', with branches
growing out of each curve.

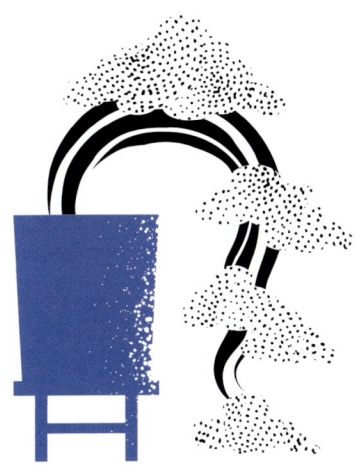

Cascada
Kenjai
A downward-bending tree,
mimicking those shaped
by heavy winter snow or
landslides in nature.

Windswept
Fukinagashi
Leaning to one side, with
all the branches growing
in a single direction.

Bonsai-in-rock
Ishisuki
Grown within the rock
itself, with roots inside
its nooks and crannies,
resembling a tree perched
atop a cliff or a mountain.

Raft
Ikadabuki
Branches sprout from a
trunk grown along the
ground, which eventually
rots away, leaving the new
trees and elevated roots.

Root Systems

Twenty to thirty per cent of a tree's biomass exists below ground level. Root depth can vary greatly by species and habitat, with trees in drier environments tending to develop deeper taproots to help them locate water and micronutrients. The desert velvet mesquite and South African shepherd's tree are true-to-form examples.

70 m (230 ft)				
60 m (197 ft)				
50 m (164 ft)				
40 m (131 ft)				
30 m (98 ft)				
20 m (65 ft)				
10 m (33 ft)				

Scots pine
Pinus sylvestris

Tree height: 35 m (115 ft)
Root depth: 33 m (108 ft)

Velvet mesquite
Prosopis velutina

Tree height: 12 m (39 ft)
Root depth: 53 m (174 ft)

Pine tree
Pinus ponderosa

Tree height: 60 m (197 ft)
Root depth: 24 m (79 ft)

Umbrella thorn
Vachellia tortilis

Tree height: 15 m (49 ft)
Root depth: 35 m (115 ft)

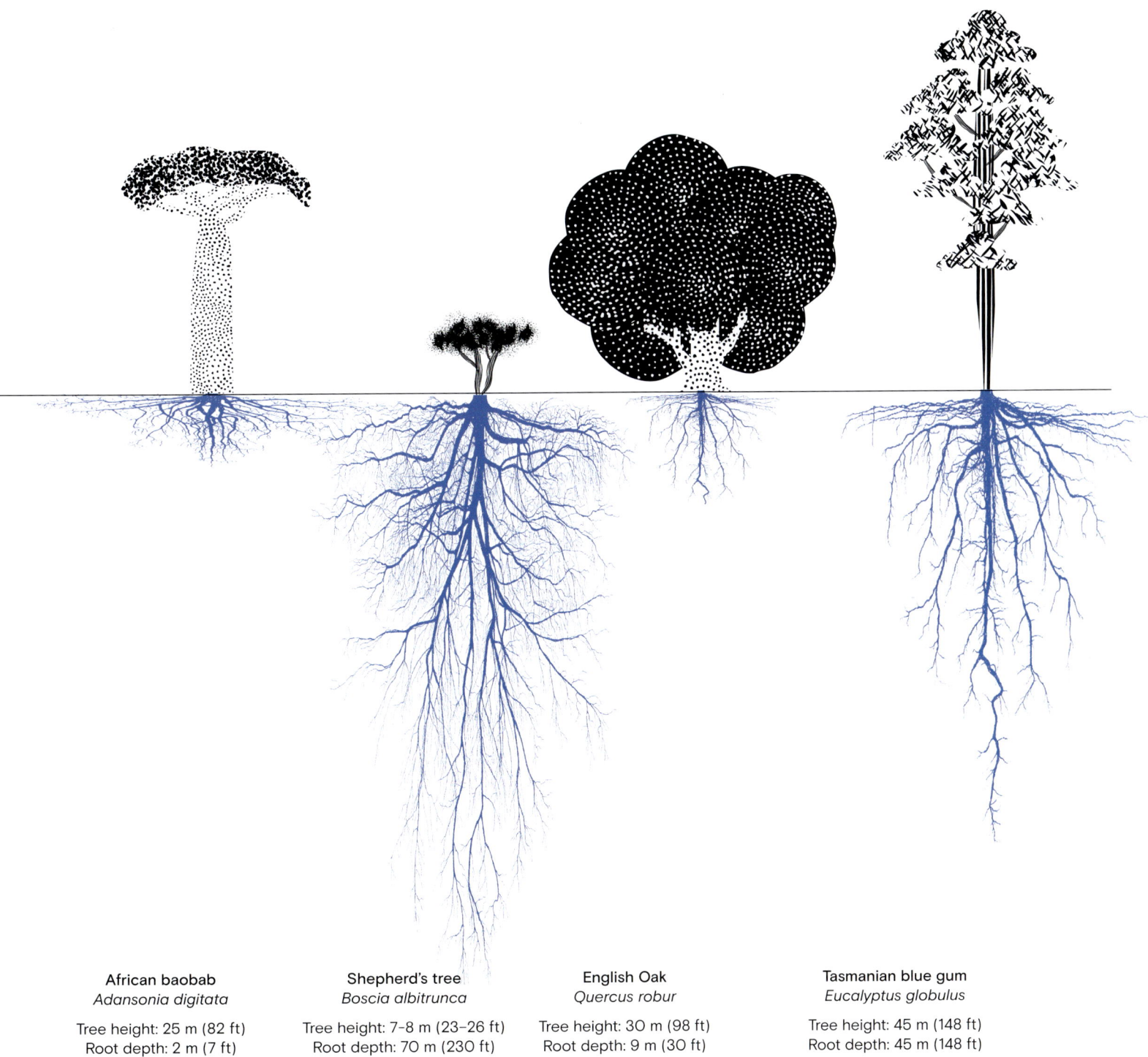

African baobab
Adansonia digitata

Tree height: 25 m (82 ft)
Root depth: 2 m (7 ft)

Shepherd's tree
Boscia albitrunca

Tree height: 7–8 m (23–26 ft)
Root depth: 70 m (230 ft)

English Oak
Quercus robur

Tree height: 30 m (98 ft)
Root depth: 9 m (30 ft)

Tasmanian blue gum
Eucalyptus globulus

Tree height: 45 m (148 ft)
Root depth: 45 m (148 ft)

Adansonia digitata, Boscia albitrunca, Quercus robur, Eucalyptus globulus

Sentience and Mythology

The ability of trees to 'communicate' with each other via electrochemical signals is comparable to neural networks and a sort of collective brain. The fact that some communication between trees precipitates the sharing of resources and early warning systems has led some (non-scientific) authors to suggest that trees are sentient beings, capable of acting rationally and communally. Such an idea is not new, of course. Folklore and mythology are richly endowed with the sentience of trees, as is children's literature [pg.266]. In many of these stories, trees and forests are of an ambiguous nature. The forests of the *Brothers Grimm* fairy tales, for example, are dark and forbidding places where witches and dwarves live. Hansel and Gretel, Snow White and Little Red Riding Hood all encounter danger in the forest, and somehow the trees seem complicit in the children losing their way.

In older European mythology, the druids spoke with trees – especially oak trees – which were thought to be endowed with powers of divination. Talking trees also appear in Eastern literature, for example in the *Shahnameh*, the epic 10th-century Persian poem, where Alexander the Great visits a talking tree. Sentient trees are different from tree spirits, which are of trees but are not the trees themselves.

> '*A wise man plants a tree under whose shade he will never sit.*'
> — *Greek proverb*

In ancient Greek mythology, the 'Dryads' were the spirits of oak trees that took the form of beautiful young women and lived as long as the trees they inhabited. In some African cultures, meanwhile, people are believed to be able to turn themselves into trees; in Zambia's Luangwa Valley I saw the shaft of a spear stuck in a tree trunk, thrown more than one hundred years before by a Nguni warrior who believed he was spearing an enemy.

Whatever the truth about the sentience of trees, there is no denying that they are impressive beings. Taller, larger, heavier and longer-lived than any other organism on the planet, they have witnessed nearly all of human-documented history, weathered centuries of storms and perhaps even acquired wisdom beyond our understanding. An old Greek proverb says that 'a wise man plants a tree under whose shade he will never sit', and the benefit of trees to generation upon generation is indisputable.

↓ Pan and Hamadryad
Eighteenth-century tile mosaic based on Roman original, Italy

A hamadryad is a Greek mythological being that lives in trees. Some maintain that a hamadryad is the spirit of the tree itself.

↘ The Shahnameh
1420–40

'Victorious king, there is a marvel here, a tree that has two separate trunks together, one of which is female and the other male, and these splendid tree limbs can speak....'

Spirituality

The tree is frequently revered as a symbol of life, wisdom and renewal, but individual specimens also hold great spiritual significance to cultures and religions the world over.

← Peepal tree
Ficus religiosa
Women walking around a sacred peepal tree during a ritual in Rajgir, India.

↓ Buddha-head statue
This Buddha-head statue at Wat Mahathat in Ayutthaya, central Thailand, has been engulfed by a *Ficus religiosa* (peepal or sacred fig tree). Native to the Indian subcontinent and also known as the bodhi tree, the *Ficus religiosa* is venerated in Hinduism, Buddhism and Jainism.

Art

Representations of trees feature strongly in art, including sculpture, their forms ranging from the elegant to the grotesque, mimicking the artist's world view.

← **Desnatureza**
Henrique Oliveira, 2011
Plywood, 310 x 380 x 360 cm (122 x 150 x 142 in.).
Galerie Vallois, Paris.

↓ **Baitogogo**
Henrique Oliveira, 2013
Plywood, 674 x 1179 x 2076 cm (108 x 464 x 817 in.).
Palais de Tokyo, Paris.

Bark

Bark

The bark of a tree is very like the skin of a human being. It is the tree's first line of defence, protecting the living tissues beneath from predators, pests, disease, fire, sunlight and unpredictable weather. It also shields the tree's cambium layer, where resin, gum and latex are produced.

01

01 – Cork oak
Quercus suber
Cork bark is highly unusual
in that it grows back after
harvesting. It takes about
nine years for the bark
to regenerate, ready for
collecting again.

Purpose and Function

Bark comprises dead, corky tissue that is impervious to water and gas, and beneath it, living cork cambium, which is normally only one cell-layer thick. Shortly after the cambium forms, some of the cells divide – this process is responsible for creating the outer layer of bark. As with human skin, underneath the bark is the tree's circulatory system, consisting of 'phloem' and 'xylem' tissue, which transports nutrients and water to all parts of the tree.

Bark's primary function is to protect the tree, providing a physical barrier against the elements, predators, pests and disease. To this end, bark is made up of a range of chemical compounds that resist physical and microbial attack and the ingress of water – including waxy 'suberins', 'cutins', 'lignins', 'tannins' and other complex macro-molecules. If injured, a tree will also produce 'terpenes' in the form of resin or polysaccharide gums. All of this adds up to an impressive cocktail of chemicals that defend the tree, but these compounds are also useful to people in various ways.

Fire Resistance and Flammability

Given the inherent combustibility of wood, perhaps the most remarkable adaptation conferred by bark is fire resistance. The best-known example of this is the cork oak (*Quercus suber*), which produces a layer of cork that can grow up to 30 cm (12 in.) thick, and which is almost impervious to fire. To test this out, try setting light to a cork from a wine bottle – not only is cork more or less impossible to set alight, but it burns extremely slowly and with almost no smoke. There are about 2.2 million hectares (5.4 million acres) of cork forest growing around the Mediterranean Basin, from which about 200,000 metric tonnes (197,000 imperial tons) of cork are harvested annually [pg.132]. Half of this bounty comes from Portugal, a country where the cork forests have increasingly been replaced by highly combustible *Eucalyptus* monocultures, favoured because they are so fast-growing.

This is both bad for the planet – cork is by far the most environmentally friendly type of wine stopper – and for people: the wildfires that swept through Portugal in June 2017, killing sixty-six people and injuring more than two hundred, are expected to occur with increasing frequency as long as people continue to plant highly combustible *Eucalyptus*. *Eucalyptus* forests were also overwhelmingly responsible for the Oakland firestorm in California in 1991, which saw the death of twenty-five people and caused an estimated US$1.5 billion of economic damage. A subsequent US National Park Service study showed that *Eucalyptus* produces nearly three times the fuel load of native oak woodland.

Eucalyptus is highly flammable due to the high concentrations of volatile oils produced by its leaves. It is native to Australia, where there are around seven hundred different species of *Eucalyptus* (or gums). In their native habitats, these trees are adapted to survive regular bushfires, either regenerating from underground 'lignotubers' (a swelling at the base of the tree) or reproducing from fruits that release their seed after fire; recent studies have shown that the germination of such seeds is often triggered by the chemicals in smoke [pg.21]. Gum trees also have a habit of shedding their highly flammable bark, adding to the fuel load on the forest floor. In the absence of regular, small-scale bushfires to clear this litter, it can cause extremely hot, fast-burning fires, as evidenced by the series of rampant wildfires that engulfed Queensland, New South Wales and Victoria in 2020. Dangers aside, the huge array of *Eucalyptus* bark textures and colours create some of the most beautiful landscapes in the world. Perhaps the most spectacular of all is the rainbow gum (*Eucalyptus deglupta*), which is native to Papua New Guinea, Indonesia and the Philippines.

> *Eucalyptus* is highly flammable due to the high concentrations of volatile oils produced by its leaves.

Keeping Animals Out

As well as the ability of some bark to withstand fire, it can play an important role in keeping browsing animals at bay. In southern Africa, for example, the knobthorn (*Acacia nigrescens*) is covered with horny, thorn-like protuberances that make it difficult for elephants to strip them of their bark [pg.122]. African elephants consume up to 300 kg (660 lb) of vegetation each day, and bark is one of their favourite food sources. Their aim is to get at the inner cambium tissue containing the phloem and sugars, but to do this they strip off both the inner and outer bark, often killing the tree in the process. The variety of complex chemical compounds found in bark – including 'cellulose', lignin, suberin, 'alkaloids', tannins, 'terpenoids' and 'saponin' – are difficult for mammals to digest, but microbes in the animal's gut help to break them down. Porcupines, giraffes, voles, badgers, deer, bears and even tree-dwelling marsupials such as koalas are also fond of bark. In addition to providing nutrients, the compounds in tree bark have valuable medicinal properties, and there is strong scientific evidence to suggest that elephants and other animals seek out specific trees for this very reason.

From the tree's perspective, however, the idea is to keep animals out, particularly when it comes to their greatest enemy – insects. There are around six thousand different species of bark beetle worldwide, which not only do terrible damage to trees themselves, but also act as vectors of other pests and diseases [pg.140]. The mountain pine beetle (*Dendroctonus ponderosae*), for example, is native to North America and over the

02

last two decades has killed millions of hectares/ acres of lodgepole (*Pinus contorta*) and ponderosa pine (*Pinus ponderosa*) in British Columbia and Colorado. Normally, bark beetles only attack and kill weakened trees, because healthy trees can repel insects by producing insecticidal gum or resin. However, unusually hot, dry summers and mild winters combined with ageing tree populations and single-species planting have contributed to the epidemic. Bark beetles lay their eggs under the bark, where the larvae feed on the cambium, disrupting and preventing the flow of nutrients and water through the phloem and xylem. In addition, bark beetles can introduce fungi to the inner bark that block water and nutrient transport and prevent the tree from producing resins to combat the beetle. The European elm bark beetle (*Scolytus multistriatus*) and the American elm bark beetle (*Hylurgopinus rufipes*) both transmit the fungus responsible for Dutch elm disease, which originated in Asia but has devastated elm trees in Europe and North America since its introduction.

02 – Elephants eating bark
Bark is a favourite food of African elephants. Here, they can be seen stripping bark from an acacia tree in Amboseli National Park, Kenya.

03 – Lodgepole pine
Pinus contorta
The top of MacDonald Pass in the Continental Divide near Helena, Montana, shows substantial damage caused by the mountain pine beetle (*Dendroctonus ponderosae*).

Antimicrobial Activity

Considering the damage caused by fungi, bacteria and other microbes to trees, it is not surprising that tree bark is full of antiseptic and antimicrobial compounds. Over the past few decades, largely due to the inadvertent human introduction of pathogens, we have seen an unprecedented rise in tree diseases caused by microbes. A high-profile example is ash dieback, caused by the fungus *Hymenoscyphus fraxineus*, first scientifically described in 2006, and currently causing up to 85 per cent mortality in the European ash (*Fraxinus excelsior*). Similarly, in the southern hemisphere, Phytophthora dieback, caused by the fungus *Phytophthora cinnamoni*, has devastated vast swathes of native Australian flora and is now present in over seventy countries around the world. It also threatens commercial crops such as avocado and ornamentals such as azaleas, camellias and boxwoods.

> Considering the damage caused by fungi, bacteria and other microbes to trees, it is not surprising that tree bark is full of antiseptic and antimicrobial compounds.

Such pathogens become a problem when they are exposed to tree species that have no natural resistance – either through the activities of humans moving infected plant material around the world or due to climate change. In the case of ash dieback, the disease originated in Asia (possibly in Japan), where native species such as Chinese ash (*F. chinensis*) and Manchurian ash (*F. mandshurica*) can tolerate infection, showing only mild symptoms on their foliage, having co-evolved with the fungus over thousands of years. Such co-evolution – essentially localized arms races between the disease and the host tree – has led to an enormous array of anti-microbial compounds in the bark of the world's trees. The vast majority of these remain unknown to science but are well understood by indigenous people and form an important part of traditional medicine pharmacopoeias.

03

Bark Medicines

One of the best-known bark medicines [pg.126] is salicylic acid, found in the bark of willow (*Salix*) trees, and the active ingredient in aspirin. Bitter infusions of willow bark were used by country herbalists for thousands of years to treat rheumatism, chills and 'the ague' (any disease characterized by a high fever). The rationale was that plants found in damp places would be effective in treating these shivery complaints. This traditional knowledge was found to have a chemical basis with the isolation of salicylic acid in the 1850s, and the synthesis of acetylsalicylic acid or 'aspirin' by Bayer in the 1890s to become the world's most widely used synthetic drug. Quinine [pg.278], used for treating malaria, and derived from the bark of several trees in the South American genus *Cinchona*, has a more interesting history. Malaria was not documented by the Aztecs or Mayans, and it is believed that the disease was brought to the New World by European settlers or the West Africans they enslaved in the 16th century. Nevertheless, the bark of *Cinchona* was used by Amerindians to treat fever due to its general 'febrifugal' (fever-relieving) properties, a remedy documented by Jesuit Brother Agostino Salumbrino (1561–1642), an apothecary who lived in Lima, Peru. He observed the Quechua Indians of Ecuador using the bark for treating chills, and found it was effective for combating malaria; *Cinchona* bark was adopted by the Jesuits for this purpose and brought to Europe in 1632.

04

As a general rule, if a medicine is used widely across a region by different tribes or ethnic groups for treating the same condition, then it is more likely to be effective.

As with willow bark, traditional medicines often utilize plants that are associated with the cause of the problem – but this doesn't guarantee a cure. In addition, some products are preventative rather than curative.

04 – Quinine
Cinchona
Handcoloured steel
engraving by Debray after
a botanical illustration by
Edouard Maubert from
*La Regne Vegetal: Flore
Medicale*, L. Guerin, Paris,
1864–71.

05 – Myrrh
Commiphora myrrha
Myrrh comes from the resin
of *Commiphora myrrha*,
a tree native to the horn
of Africa and Arabia.

05

As a general rule, if a medicine is used widely across a region by different tribes or ethnic groups for treating the same condition, then it is more likely to be effective. If, on the other hand, a species has a highly localized use associated with the beliefs of a particular group, it is less likely to have a proven purpose. For example, the bark of the tree *Cassia abbreviata* is used in traditional medicine across south-central Africa to terminate pregnancies or to induce labour, indicating that it is probably highly effective.

Some traditional medicines are extremely toxic if taken in the wrong way or at too high a dose, and may be used as a poison and/or in relation to witchcraft. An example of the latter is the tree bark of *Erythrophleum sauveolens* – the African ordeal tree – which contains the alkaloid 'erythrophlein', a cardiac depressant that causes heart failure. In the past, a decoction of the bark was used to 'identify' witches during a public trial in which the accused was made to drink the poison in full view of the community. If the person vomited up the mixture, they were deemed to be innocent; if not, they would die from the poisoning or at the hands of their peers. The outcome of such trials was not as random as may be supposed, because the higher the dose, the more likely you'd be to vomit – leaving lives very much in the hands of the witchdoctor administering the mixture. *Erythrophleum sauveolens* is one of hundreds of African tree species that are used as fish poison. Another is the closely related red syringa (*Burkea africana*), the bark of which is pounded into a paste or powder and dropped into a pool of water to stupefy the fish, which can then be harvested as they float to the surface. Today, these practices are generally outlawed because they don't discriminate between mature and young fish, with high doses killing all the smaller individuals and wiping out entire populations – clearly an unsustainable way to fish.

Other Uses of Bark

Bark has been collected for millennia to produce dyes [pg.130] for textiles. Alder buckthorn (*Rhamnus frangula*) bark has been used in this way since the Iron Age, giving colours ranging from mustard yellow through to cinnamon red. The dyeing process typically involves soaking the bark in very hot water, adding soda ash and allowing the mixture to ferment over several weeks or months, stirring and topping up with water and soda ash to keep the pH high. The fabric is then steeped in the strained liquid for two weeks. Fermentation dyeing works well with barks of plants rich in tannin (for example oak, maple, willow and birch), which give a range of brown and yellow dyes.

In the past – and to a certain extent still today – bark was used to make cloth itself. The paper mulberry (*Broussonetia papyrifera*), for example, is indigenous to sub-tropical Asia, and was used by ancient Austronesians for this purpose. Barkcloth is made by beating sodden strips of the fibrous inner bark of these trees into sheets, which are then finished into items of clothing. Cultivation of the paper mulberry follows Austronesian migration patterns from 5000 to 500 BCE, originating in East Asia and spreading as far as Papua New Guinea and Oceania. Barkcloth is widely known as 'tapa' cloth, a name derived from Tahiti and the Cook Islands, but now in common usage internationally with various regional derivations. In Hawai'i it is called 'kapa' while in Madagascar, which was colonized by Austronesians during the period 350 BCE

to 550 CE, the term 'tapia' is applied to cloth that is made from silkworms that feed on the tapia tree (*Uapaca bojeri*). In Africa, barkcloth is made from a number of different tree species, including the Natal fig (*Ficus natalensis*, a relation of the paper mulberry), the baobab (*Adansonia digitata*) and various species of the genus *Brachystegia* – the main constituent of south-central Africa's miombo woodlands. Barkcloth production generally does not involve weaving, but the fibres produced from bark can be used to make a wide range of other products, including fishing nets, yarn and rope. In Papua New Guinea, the 'bilum' (or 'noken') is a string bag traditionally woven from tree bark, in which children are carried; it has deep cultural significance as well as being a useful accoutrement.

Bast fibres, derived from the inner bark of trees such as lime, wisteria and mulberry, were once used to make string and yarn, and some of these natural fibres are still put to this purpose today. Jute, which is used to make hessian and burlap, comes from the plants of the genus *Corchorus*, which although not trees, can grow several metres in height. Modern industrial papermaking uses cellulose fibres extracted from wood pulp, but traditional papermaking employs bark fibres. In the Himalayas, paper is still made from the bark of the lokta tree (*Daphne papyracea*) in the plant family *Thymeleaceae*, while in Madagascar, the avoha tree (*Gnidia daphnifolia*), which belongs to the same family of plants, is the source of bark fibres harvested to make the country's famous Antemoro paper, embedded with wildflowers.

Persimmon tree
Diospyros virginiana
The thick, crocodile-skin bark of the persimmon helps to protect the tree against plummeting winter temperatures, as well as drying winds and wood-eating pests.

Adaptations

The different properties of tree bark help species adapt to their unique habitats. Obtrusive thorns ward off hungry herbivores, while ill-tasting or toxic chemicals and resin perform the same function against fungi and insects.

Red syringa – *Burkea africana, Chemical*
People use the chewed bark as a poultice on septic sores.

Baobab – *Adansonia digitata, Physical*
Baobab bark is thick and spongy, and can grow back after damage.

Cork oak – *Quercus suber, Physical*
Cork oak bark is a natural fire retardant, and protects the tree from forest fires.

Knobthorn – *Acacia nigrescens, Physical*
The thick, woody bosses of the knobthorn help prevent elephants from stripping the bark.

Crystal bark – *Crossopteryx febrifuga, Physical/chemical*
This savannah tree has hard silicon crystals embedded in its bark to deter browsers, as well as antimicrobial chemicals.

Rainbow eucalyptus – *Eucalyptus deglupta, Physical*
The stringy, flaking bark of many Eucalyptus species is shed when it catches fire.

Frankincense – *Boswellia sacra, Physical/chemical*
Gums and resins form a physical and chemical defence against infection if a tree is wounded.

Colour

Tree bark comes in a range of colours; most commonly reddish brown, it can also take on shades of green, red, orange, grey and white. More exuberant still are striped or multi-coloured barks.

← Rainbow eucalyptus
Eucalyptus deglupta
The rainbow gum is a riot of colour; young bark is bright green, while older layers turn shades of dark green, rust red, purple and orange.

↓ Wood for Trees (Camo)
Charlotte Evans, 2016
Gouache on paper.

Medicines and Uses

Bark and its compounds contribute to a number
of life-saving medicines, including aspirin and
quinine – the latter is used to treat malaria.
Other uses for bark include making cloth, dye
and paper.

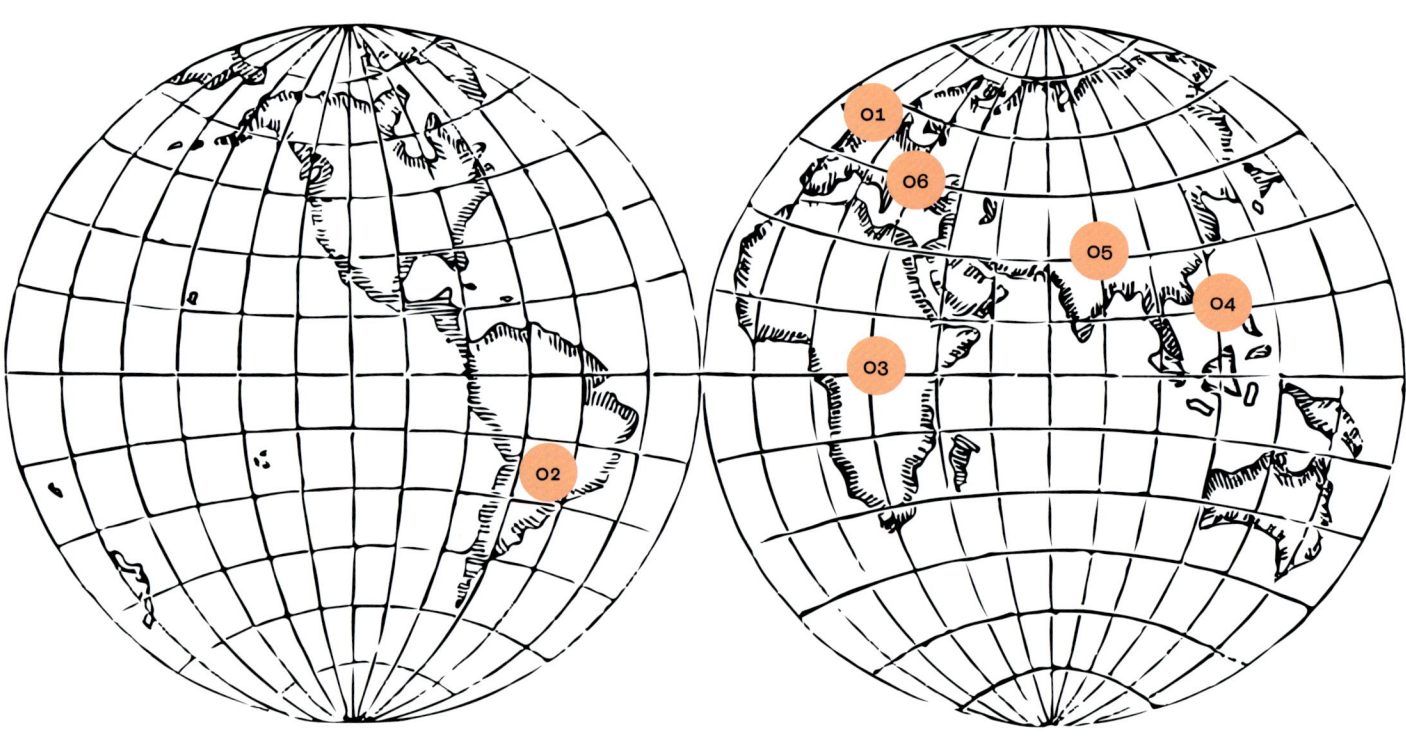

01 – Willow
Salix

Salicylic acid, derived from the bark of willow trees, is best known as the active ingredient in aspirin.

02 – Cinchona
Cinchona

Containing quinine, the bark of several trees of the South American genus *Cinchona* has long been used to treat fever; quinine is still used to treat malaria today.

03 – Long-tail cassia
Cassia abbreviata

The bark of the long-tail cassia is used across swathes of south-central Africa to terminate pregnancies.

04 – Paper mulberry
Broussonetia papyrifera

By pounding the fibrous inner bark into sheets, the paper mulberry was harvested by the Austronesians to make barkcloth.

05 – Lokta
Daphne papyracea

Up in the Himalayas, paper is made using the bark of the lokta tree.

06 – Alder buckthorn
Rhamnus frangula

Alder buckthorn bark has been used to produce a yellow, orange or red dye since the Iron Age.

Texture

Bark can be thick or thin, smooth or cracked, knotted or uniform. Different bark textures and patterns are excellent sources of inspiration for artists, designers and craftspeople.

01

02

03

01 – Tree Bark
Brett Weston, c. 1950
Silver gelatin print.

03 – Tree Bark
Brett Weston, c. 1970
Silver gelatin print.

05 – Tree Bark
Brett Weston, c. 1975
Silver gelatin print.

02 – Bark, Europe
Brett Weston, c. 1971
Silver gelatin print.

04 – Tree Bark
Brett Weston, 1977
Silver gelatin print.

06 – Bark Abstraction, Europe
Brett Weston, 1971
Silver gelatin print.

04

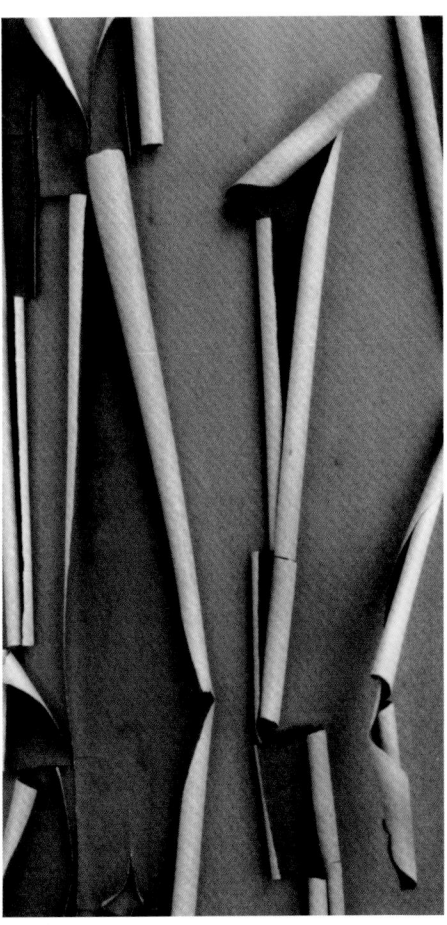

05

06

Dyes

Bark has been used for millennia to create
coloured dyes for a range of textiles.
Results can be surprisingly bright and vibrant.

Black oak
Quercus velutina

Alder buckthorn
Frangula alnus

Willow
Salix

Coloured wool

Different wools created
using natural dyes,
ranging from off-white
to rich orange.

Silver birch
Betula pendula

Sumac
Rhus glabra

Poplar
Populus

Coloured wool
Different wools created
using natural dyes,
ranging from off-white
to rich orange.

Silver birch
Betula pendula

Sumac
Rhus glabra

Poplar
Populus

Bark ⟶ Dyes ⟶ Quercus velutina, Frangula alnus, Salix, Betula pendula, Rhus glabra, Populus

Cork

One of the most notable characteristics of cork is its natural fire resistance. Bark from the evergreen cork oak is collected for wine-bottle stoppers, cork flooring and insulation. Because bark can be harvested without killing the tree, cork is an important sustainable resource.

← Cork tree
Quercus suber
The cork oak produces a layer of cork that can grow up to 30 cm (12 in.) thick. Humans have been using cork for more than 5,000 years; today, around 50 per cent of the world's cork comes from Portugal.

↓ Vitra Cork Family side table/stool, Model D
Jasper Morrison, 2004
Made entirely of cork, the pieces in Jasper Morrison's Cork Family use clear, geometrical lines to provide a pleasing contrast to the natural surface of the material.

Resins, Gums and Latex

Although not strictly bark products, gums, latex and resins are derived from the cambium layer just below the bark, and provide us with a huge range of useful products, from tar and lacquer to maple syrup, palm sugar and frankincense. Resins are produced by specialized cells in response to injury or attack, and they primarily comprise 'terpenes' – complex organic compounds that are insoluble in water and harden in the air. Resins provide us with tar, camphor, creosote, gamboge, lacquer, mastic, pitch, rosin, turpentine and varnishes. They also include frankincense and myrrh, well known to Christians as the precious gifts offered to the baby Jesus by the wise men from the East. Frankincense comes from the resin of trees in the genus *Boswellia*, which grow in the dry regions of Africa, Arabia and South Asia. Five species of *Boswellia* produce frank-incense. *Boswellia papyracea*, so called because of its papery bark, is native to Ethiopia and the horn of Africa, and is of great cultural significance because its frankincense is used in Coptic Church ceremonies and Ethiopia's coffee-making ritual. Due to a combination of climate change and over-harvesting, this species is becoming scarcer, leading to shortages of frankincense, which has traditionally only been harvested from wild trees. Myrrh comes from a closely related species, *Commiphora myrrha*, which is native to the horn of Africa and Arabia. Like frankincense, myrrh has important religious significance. According to the Gospel of St John (19:39), Joseph of Arimathea and Nicodemus bought a mixture of myrrh and aloes in which to wrap Jesus's body. Joseph of Arimathea has his own tree-related legend, concerning the original Glastonbury thorn, which reputedly sprouted when he thrust his staff into the ground while visiting Glastonbury, England with the Holy Grail [pg.184].

Unlike resins, gums and latex are derived from tree sap, and are made up of chains of chemical compounds known as 'polysaccharides' and 'polymers'. They include sugary saps such as birch syrup, maple syrup [pg.298] and palm sugar, but also gums such as chicle and gum Arabic. Chicle is a natural gum that is collected from several central American trees of the genus *Manilkara*, including *Manilkara chicle* itself. Traditionally used by the Aztecs and Maya, it was chewed to assuage hunger, freshen breath and keep the teeth clean. Chicle was commercialized in the 1860s when it was brought to New York from Mexico by the former Mexican President, Antonio Lopez de Santa Anna; there, he gave it to entrepreneur Thomas Adams, who marketed it as Adams New York Chewing Gum in 1871. Chiclets and Wrigley's spearmint gums became popular at the end of the 19th century and are still in production today. However, by the 1960s chewing-gum makers had switched to butadiene-based synthetic rubber because chicle production was unable to match market demand. Gum Arabic, meanwhile, is a general term used to describe natural gums derived from tree species such as *Acacia senegal* and *Acacia seyal*, which are indigenous to North Africa and the Sahel. Gum Arabic is both edible and soluble in water, and is a mixture of proteins and complex sugars. It is used in the food and drink industry mainly as a stabilizer, but also in printing, cosmetics, glue and paint production. In 2019, total gum Arabic exports were estimated at 160,000 metric tonnes (157,000 imperial tons), almost entirely harvested by tapping wild trees and constituting an important income source for Sahelian countries in Africa.

Rubber tree
Hevea brasiliensis
Collecting rubber, used in
the production of tyres,
clothing, gloves, tubes and
hoses, paint and toys.

Latex, like resin, is produced as a defence against attack or injury and is a complex emulsion – consisting of water, proteins, starches, sugars, alkaloids, oils, tannins and gums – that coagulate on exposure to air. Latex is found in particular families of plant, including the spurge family (*Euphorbiaceae*), the dogbane family (*Apocynaceae*) and the fig family (*Moraceae*). Natural latex is usually milky white, but some plants produce yellow, orange or even scarlet latex. Probably the best-known latex-producing tree species is the rubber tree – *Hevea brasiliensis*. As its scientific name suggests, *Hevea* is native to the Brazilian Amazon and surrounding countries where, in its natural state, it can grow to over 40 metres (140 feet) tall. Its milky-white or yellow latex is produced by latex vessels found in the bark outside the phloem, which spiral up around the trunk in a right-handed helix at an angle of about thirty degrees. Latex is extracted by humans by tapping the tree: incisions are made across the latex vessels from where the drips are collected in a container fixed to the tree. Although tapping doesn't harm the tree, it does reduce its growth rate, so plantation trees tend to be much smaller than those found in the wild. In addition, latex production is greatest in trees between ten and thirty years old, after which it declines, and trees are cut down.

Raw latex is rather soft and unstable. It is made into rubber through vulcanization, which involves mixing latex with sulphur and heating it until it hardens. This process, discovered accidentally by Charles Goodyear in 1839, led to the industrialization of rubber and its use in a wide range of products including, most famously, car tyres.

> The more tree species we push to extinction, the fewer options future generations will have to access these valuable natural products.

Although the nascent rubber industry was centred around Belem, Santarem and Manaus in Brazil for much of the 19th century, rubber plantations were established by British colonists in India, Malaya (Malaysia) and Ceylon (Sri Lanka) by the 1890s; today the main rubber-producing nations are Thailand, Indonesia, Malaysia, India, China and Vietnam. This is in part because rubber trees in the Amazon are susceptible to South American leaf blight, caused by a fungus native to the Amazon – *Microcyclus ulei*. In 2017, annual global rubber production was 28 million metric tonnes (27.5 million imperial tons), around half of which was natural rubber and the remainder synthetic rubber derived from petroleum.

From an environmental perspective, natural rubber might seem a more sustainable alternative to rubber from fossil fuels. However, unfortunately in a number of countries across Asia, large areas of natural rainforest have been cleared to make space for rubber plantations, reducing sequestered carbon and devastating biodiversity. In addition, given that Asian rubber plantations have been established from a very narrow gene pool of *Hevea* cultivars that are susceptible to South American leaf blight, it is probably only a matter of time before the disease finds its way to Asia and obliterates crops there. In fact, there are around twenty thousand plant species from more than forty plant families that produce latex, some of them with the potential to produce commercial rubber. However, the more tree species we push to extinction, the fewer options future generations will have to access these valuable natural products.

Frankincense

Extracted from the genus *Boswellia*, frankincense comes from resin produced below the bark and is used in incense, perfumes, soaps and lotions. These trees favour arid climates and are found across the dry regions of Africa, Arabia and South Asia. In Ethiopia, frankincense is central to both Coptic Church rituals and traditional coffee ceremonies.

01

01 – Frankincense tree
Boswellia sacra
Frankincense trees are increasingly threatened in the wild by climate change and grazing livestock.

02 – Nineteenth-century botanical print of *Boswellia carterii*, a scientific synonym for *Boswellia sacra*.

03 – Frankincense is derived from the milky resin of *Boswellia*, exuded when the bark is slashed.

04 – *Boswellia* gum hardens into the crystalline substance that is frankincense.

05 – Used since ancient times, frankincense is known as *al-libān* or *al-bakhūr* in Arabic.

03

02

04

05

Architecture

Incorporating tree bark into external façades and interior walls can add texture to modern architecture. It's also a sustainable choice, given the regenerative properties of bark.

↓ **Wooden house, Switzerland**
Atelier Risi, 2020
Located in Böschi, this wooden family house was designed by Atelier Risi and is wrapped in a treated tree-bark façade. The entire house was built using biodegradable materials and timber from the nearby Ägeri Valley.

→ **Christian Louboutin store, Miami**
212box, 2017
Tree panels cover the exterior of Christian Louboutin's two-storey flagship store in Miami. The textured design is replicated inside, where the walls are lined with white birch, gold birch and pin cherry wood.

Tree Damage

There are some 6,000 different species of bark beetle, and though the majority target dead or dying trees, the epidemic has affected healthy ones, too. Bark beetles can do real damage, as their larvae feed on the nutrient-rich inner bark (phloem). Several adaptations – including the release of chemicals and sap – serve as a tree's first line of defence [pg. 122].

← Traces of beetles under the bark on a spruce tree (*Picea*)

After the beetle has laid its eggs beneath the bark layer, the emerging larvae then feed on the cambium – which hinders the flow of nutrients and water to the tree.

↘ Tree casualties

Bark beetles have killed hundreds of millions of conifers in North America and Asia over recent decades. Climate change has exacerbated the impacts of beetle damage, often due to drought stress in host trees.

Wood

Wood

It is hard to imagine life without wood. I am sitting in my office at a wooden desk with wooden bookshelves and other wooden furniture arranged on a wooden floor. If all of this was made of stone or metal, the world would be a cold place to the touch and to the eye.

01

01 – Spruce tree
Picea
The flowing contours of
wooden veneers reflect
the plane in which the
wood is cut across
the tree's growth rings.

Wood is my favourite material: it's not too hard and not too soft, and retains that essence of organic nature. My grandfather was a piano-maker, and I grew up with the smell of his workshop and his gentle coaching on working with wood ('let the tools do the work'), so it's in my blood. Despite this, I take wood for granted, rarely thinking about where it came from and the journey it took to reach this room.

My desk, which is made from English oak (*Quercus robur*), started its life as an acorn. Initially its stalk was composed solely of living tissue, but as it matured, that living tissue lignified into wood that, over hundreds of years, accumulated the quantities desired for furniture-making. If my oak tree was 1.4 metres (4.5 feet) in diameter when it was cut down, it must have sprouted around the time of the Battle of Waterloo, some two hundred years ago.

Looking at a cross-section of a tree trunk reveals five distinct layers of tissue – bark, 'phloem', 'cambium', 'sapwood' and 'heartwood'. The bark [pg.110] is a protective layer made up of largely dead tissue. Immediately inside the bark is the phloem tissue, responsible for transporting sugars and nutrients around the plant. Next is the cambium layer, comprising actively dividing cells, and within it, the sapwood, through which water and minerals are transported from the roots to the rest of the tree. Finally, the heartwood is the central pillar of the tree, made of dead cells that are reinforced with lignins, phenols, resins and other organic compounds that give the wood strength and protect it from insects and microbes.

If my oak tree was 1.4 metres (4.5 feet) in diameter when it was cut down, it must have sprouted around the time of the Battle of Waterloo, some two hundred years ago.

The woodsman who cut down my oak tree would have been able to calculate its age pretty accurately by counting the growth rings of the heartwood and sapwood of the tree: one ring for each year of its life. The science of ageing a tree by counting its growth rings is called 'dendrochronology', and because growth rings vary in size depending on the climatic conditions of any given year in any given location, it is often possible to identify both the date and place that a particular wood was harvested. For example, Henry VIII's famous warship, the *Mary Rose*, which sank in the Solent on 19 July 1545, was made from oak timbers harvested from East Anglia in 1510–11; dendrochronology records have helped distinguish between original boat timbers and those that were later refitted. In some areas, records for certain species of tree go back thousands of years – in the case of the oldest tree species, the bristlecone pine [pg.86], more than ten thousand years. The biggest gaps in dendrochronology records in northern Europe coincide with the advent of the Black Death in the 14th century, when there was a hiatus in building caused by the bubonic plague, which wiped out tens of millions of people.

Both the sapwood and the heartwood are used for timber, though the sapwood is not as durable as the heartwood. It is also lighter in colour. For cabinetmakers selecting wood for furniture, colour [pg.162], texture, durability and malleability are all important considerations. Types of wood that are almost pure white include sycamore (*Acer pseudoplatanus*), spruce (*Picea*) and holly (*Ilex*). Yellow woods are derived from some pine species (e.g. *Pinus ponderosa*) and boxwood (*Buxus sempervirens*), while red woods include red pine (*Pinus resinosa*), sequoia (*Sequoia sempervirens*), mahogany (*Swietenia*) and rosewood (*Dalbergia*). Brown woods come from species such as satin walnut (*Liquidambar*), chestnut (*Castanea*) and snakewood (*Brosimum guianense*). Black woods include the ebonies (*Diospyros*) and zebrawood (*Dalbergia melanoxylon*).

Heartwood and sapwood, as layers of the same tree, should not be confused with hardwoods and softwoods [pg.164]. The latter describes the contrasting characteristics of wood from distinct tree species. The difference between hardwoods and softwoods is not just the durability of the wood, as their name suggests. Hardwoods are derived from deciduous, broad-leaf trees like oak, maple, teak and mahogany, while softwoods come from evergreen conifers (such as pine, spruce and fir). Hardwoods are usually more durable because they are denser and grow more slowly. For this reason, they are used in joinery, furniture-making, wood flooring and fine veneers. The hardness of wood can be measured using a standard 'Janka hardness test', which quantifies the force required to embed a 11.13 mm (⁷⁄₁₆ in.)-diameter steel ball halfway into the surface of the wood. In contrast, softwoods are primarily used for paper-making, plywood, panels and interior mouldings and construction.

Wood has underpinned a number of major technological advances since the very earliest hominids walked this Earth.

Wood is used to create endless products that are invaluable to humans. The timber trade – on which the construction and furniture industries depend, as well as many local artists and artisans around the world – is a thriving business, and was valued at around US$244 billion in 2019. However, the properties of wood have been useful to humans for much longer than it has been a global commodity – as a material and resource, wood has underpinned a number of major technological advances since the very earliest hominids walked this Earth [pg.158].

02

03

Wood and Human Technological Advances

It is thought that the first hominids sheltered in trees, which goes some way to explaining our long affinity with these plants. Perhaps the greatest development made by early humans in which wood played a key part was controlling fire, a skill mastered between 1 and 2 million years ago. The combustibility and calorific value of firewood made it choice fuel, providing a source of both warmth and light. The friction caused by rubbing a hardwood and softwood together means that wood can also be used to start a fire; while it is likely that fire-starting was a later development in human history, some cultures still use wood for this purpose today. Firewood not only conferred heat and light, it also meant early humans could cook food, which in turn facilitated a much wider and healthier diet.

The construction of wooden shelters some four hundred thousand years ago followed the discovery of firewood and represented another significant advance as humans moved out of caves and natural shelters. At its most basic, wood is simply a building material used for rafters, joists, flooring, roofs and walls. It has always been crucial, however, to select the right timber, with all the necessary properties for the job at hand. Alder (*Alnus glutinosa*), for example, is a light, soft and flexible wood that is water resistant and, since Roman times, has been used to make canals, water pipes and the foundations for buildings in waterlogged soils. In fact, alder itself grows in waterlogged soils, and it's no coincidence that the city of Venice is built on alder piles. Another timber almost impervious to rot is Mulanje cedar (*Widdringtonia whytei*), the national tree of Malawi, which has similar properties: mountain huts built in the 1920s at an altitude of more than 2,000 metres (6,500 feet), and exposed to more than a metre (39 in.) of annual rainfall, are still in perfect condition a century later. A few years ago, one of these huts burnt down and, because Mulanje cedar is now so scarce, was rebuilt using European cedar (*Cedrus*) – the wood barely lasted ten years before it rotted away.

Wood has been an invaluable building material through the ages. Today, its application becomes most interesting when architects working with wood take inspiration from tree form itself. A structure that embodies this philosophy is the Wayfarer's Chapel [pg.154], which is situated on a knoll on the Palo Verdes Peninsula overlooking the Pacific Ocean on the Californian coast. The Chapel – also known as the Tree Chapel or the Glass Church – was designed and built by architect Lloyd Wright, son of the more famous Frank Lloyd Wright. Completed in 1951, the Chapel was inspired by the cathedral-like majesty of California's redwood trees that surround the perimeter. The building itself is a soaring structure made almost entirely of wood and glass, with orchids, ferns and other plants growing out of the short retaining wall around it. I was lucky enough to visit a few years ago, and have never experienced anything quite like it. The redwoods that edge the Chapel provide welcome shade, while 6.5 mm (quarter in.) glass is all that separates visitors from the outside, affording a great sense of peace and quiet.

Another set of fundamental human advances made possible by wood relates to travel and transport. Stonehenge's inner circle of blue standing stones, weighing 2–4 metric tonnes (1.9–3.9 imperial tons) each, originated in South Wales, some 225 km (140 miles) from the site of Stonehenge, while the outer circle of 25 metric-tonne (24.5 imperial-ton) stones came from 24 km (15 miles) away. It is thought that the stones were dragged to the current site of Stonehenge on wooden rollers or sledges about five thousand years ago.

The first wheels appear in history in the Copper Age (4500–3300 BCE) and coincide with the domestication of the horse; wheel technology was shared between the Near East and Europe in around 3500 BCE.

Early wheels were solid wooden affairs – initially just cross sections of tree trunks, but then later made of rounded planks. Here the tensile strength of the wood would have been important: wood that cracks or splits easily would not have been suitable for wheel-making. Spoked wheels first appeared in about 2000 BCE in the Western Asian steppe and then the horse-drawn chariots of the Caucasus. Wooden spoked wheels were the most technologically advanced wheels until the first metal wheels arrived with the railways in the 18th century, while pneumatic rubber wheels were not introduced until the 1870s. The great 19th-century European migrations in North America, Southern Africa and Australia were carried out in wooden, ox- or horse-drawn wagons. For example, the heavy Conestoga wagons of the American pioneers were made out of hardwoods like oak and poplar, which are durable in all weathers.

Boats go back much further than wheels. The oldest-known boat is the Pesse canoe, a dugout made from the tree trunk of Scots pine (*Pinus sylvestris*) that dates from about 8000 BCE, unearthed in modern-day Holland. Although this is the oldest archaeological record of a boat, it is thought that the use of watercraft goes back to prehistoric times. Much earlier migrations in Southeast Asia, and the first human settlement of Australia forty thousand years ago, almost certainly involved travel by boat, possibly on wooden rafts or in dugout canoes. The buoyancy and water resistance of the wood are key properties for boat-making, with lighter, more resinous varieties being favoured. Again, high tensile strength is desirable, too: you don't want your boat to split open mid-ocean. As with wooden wheels, wooden ships were the zenith of boat-building technology for thousands of years until the first iron-hulled ships were constructed in the mid-19th century. One of the most famous wooden ships in the world is still in commission and can be visited at Portsmouth in England. Admiral Lord Nelson's flagship, HMS *Victory*, represented state-of-the-art shipbuilding when she was launched in 1765. The vessel is made from more than five thousand oak trees, with fir and spruce used for the decks, masts and yardarms; oak wood was selected for its strength, and fir and spruce for their lightness and flexibility. HMS *Victory*'s oak hull was up to 60 cm (2 ft) thick, and designed to resist a 20 kg (42 lb) cannonball.

The technological advances required to make wooden wheels and boats extended into many other areas of human endeavour – from spinning wheels to water wheels, windmills and propellers. The shaping and fashioning of wood, including bending it with the help of steam, has led to the development of a huge range of tools, weapons, furniture, musical instruments and handicrafts [pg.290]. Whereas throughout most of human history our insatiable appetite for wood has led to its over-exploitation, new priorities might mean it has more value left in the ground.

Carbon Sequestration, Logging and Reforestation

As discussed in Chapter 9 [pg.270], wood is an extremely effective carbon sink, making it invaluable for absorbing carbon from the atmosphere and mitigating global climate change. Scientists quantify the carbon sequestered by trees by measuring or estimating the dry weight of the tree, then dividing it by two to give the weight of carbon. Of course, it is not practical to cut up, dry and weigh a large number of trees, so the formulae developed for calculating a tree's dry weight (or 'biomass') are based on extrapolation from simple measures such as the diameter of the tree at breast height. Different species of tree grow at different rates, however, depending on factors including location and the density of their wood. This means that they capture carbon at different rates too, so ideally we need formulae for each of the world's sixty thousand tree species. It is also not as simple as just cutting the tree down, chopping it up, drying the wood and weighing it.

> Early wheels were solid wooden affairs – initially just cross sections of tree trunks, but then later made of rounded planks.

04 – Bayeux Tapestry
c. 1070
Wool embroidery on linen. Trees are chopped down to build ships for the Norman fleet.

05 – Hardwood timber
Timber harvested from Asia's hardwood forests is not being replenished by tree-planting due to the time it takes for these trees to grow.

04

05

As can be seen in Chapter 3 [pg.78], some trees have more biomass below ground than above. Due to these difficulties in measuring biomass and carbon precisely, our estimates are fairly rough and vary between sites.

Despite the importance of trees in off-setting carbon emissions and combating climate change, the world's forests are being cut down at an alarming rate. Given the number of uses humankind has for timber, it's hardly surprising that there's severe pressure on wood as a resource. An assessment of logging in the tropics published in the recent 'State of the World's Trees' report indicated that 20 per cent of the tropical forest biome was either actively logged or allocated to logging concessions between 2000 and 2005. About half of this area had already lost more than 50 per cent of its potential forest cover. Furthermore, approximately 300 million cubic metres (10 billion cubic feet) of tropical hardwood timber is harvested annually, equivalent to an estimated 100 million trees. These trees are not being replenished, which means that the price of hardwoods is only likely to increase, and that tropical hardwood species will continue to become scarcer. Ultimately, these trees will be commercially unviable to harvest or even extinct. Even if policymakers get their acts together, and world leaders stick by their promise made at COP26 to end and reverse deforestation by 2030, much of the logging is illegal. INTERPOL estimates the value of forestry crimes – including corporate offences and illegal logging – at US$51–152 billion per year. Natural forests can be managed sustainably, as evidenced by boreal forests in the north, which are actually increasing in size. In the tropics, however, we have a long way to go before we can reach this kind of equilibrium.

As with most things, there are complexities associated with tree conservation and planting that need to be taken into account. Old growth forests, for example, sequester more carbon than new plantations, so protecting these is a no brainer in carbon terms. However, the timber in old growth forests (and sometimes the land itself) is usually worth considerably more in monetary terms than the carbon, so it is important that governments either legislate against cutting down old forests or provide incentives to their citizens to preserve them for carbon-capture purposes. Countries like Costa Rica have done both, paying landowners and communities to leave forests intact, but in most places around the world governments have been slow to protect forests and even to remove perverse incentives such as granting licences to harvest timber.

> Everyone agrees that we need to plant more trees, but the question of what to plant and where is much more complicated.

Another comparatively efficient method of capturing carbon and enabling biodiversity to recover is to protect degraded forests and allow them to regenerate naturally from seeds in the soil. Where this works – usually by keeping people and livestock off the land – it isn't necessary to plant trees at all. Nature will do the job on its own. However, in many places the soil seed bank is so diminished that only weeds and invasive species emerge, causing more problems. In such cases, assisted regeneration may be necessary, mixing tree planting with the removal of weeds and invasive species. The last resort is planting trees, but this is expensive – particularly if you want to leave them in the ground – so financial incentives (or disincentives) are essential.

Everyone agrees that we need to plant more trees, but the question of what to plant and where is much more complicated. Forestry infrastructures, including seed banks, nurseries and tree-growing skills, are primarily set up for production forestry. Production forestry involves the use of genetically improved, usually exotic, species that grow fast and straight. In the northern hemisphere, species such as the Sitka spruce (*Picea sitchensis*), Douglas fir (*Pseudotsuga menziesii*), larch (*Larix*) and various pine (*Pinus*) species are commonly grown in plantation forestry. In the southern hemisphere and the tropics, *Eucalyptus*, *Acacia* and pines such as *Pinus radiata* and *Pinus patula* are widely used. All of these species are fast-growing softwoods suited to production forestry but not necessarily to carbon capture. If these trees are planted in the wrong place, for example in peatlands and some grasslands, they can actually displace more carbon than they capture because these natural habitats are more effective carbon sinks than the trees that replace them. In virtually all cases – because they are non-native species planted as monocultures – they create sterile landscapes, disrupting native biodiversity. There are other problems with monocultures too; increasingly they are proving highly susceptible to emerging pests and diseases. For example, Japanese larch (*Larix kaempferi*) plantations in northern Europe have been hit very hard over the past decade by the fungal pathogen *Phytophthora ramorum*, which has decimated larch production.

What all this means in practice is that we are not well prepared to replenish and restore the sixty thousand species of tree we find in nature. Indeed, we currently have the processes, infrastructures and skills in place to plant perhaps five hundred of them. The Bonn Challenge is a global goal to bring 350 million hectares (865 million acres) of degraded and deforested landscapes into restoration by 2030. At least half of the pledges made so far comprise plantation forestry, which at that scale has the potential to be yet another disastrous human intervention for native species and biodiversity. It is essential that we protect natural forests and other habitats and restore them with native species – only planting exotic trees in completely degraded landscapes where the negative impacts will be negligible. Without this foresight, our laudable efforts to capture carbon will simply create another problem by devastating biodiversity.

Cross-section of a tree trunk
The growth rings not only indicate the age of the tree, but also give clues about the years in which it was alive.

Wayfarer's Chapel

The Wayfarer's Chapel in California takes nature and tree form as its very inspiration. It was designed by Lloyd Wright based on the vision of a local woman, Elizabeth Sewall Schellenberg, who lived on the Palos Verdes Peninsula in the late 1920s.

← Wayfarer's Chapel,
California
Lloyd Wright, 1951
Vast redwoods – in which
Lloyd Wright saw peace,
beauty and shelter,
and which influenced
the chapel's design –
surround the building,
providing dappled shade
and enhancing the
connection with nature.

↘ The separation between
inside and outside is blurred
as the construction is made
almost exclusively of glass
and wood.

Trunk Layers

The cross section of a trunk comprises the bark, cambium, sapwood, heartwood and pith. The depth of the different layers depends on the species, the density of the wood and how fast the tree grows.

← **Trunk variation**

The colours and textures
of trunk layers vary from
species to species and even
from tree to tree. Every piece
of wood is unique.

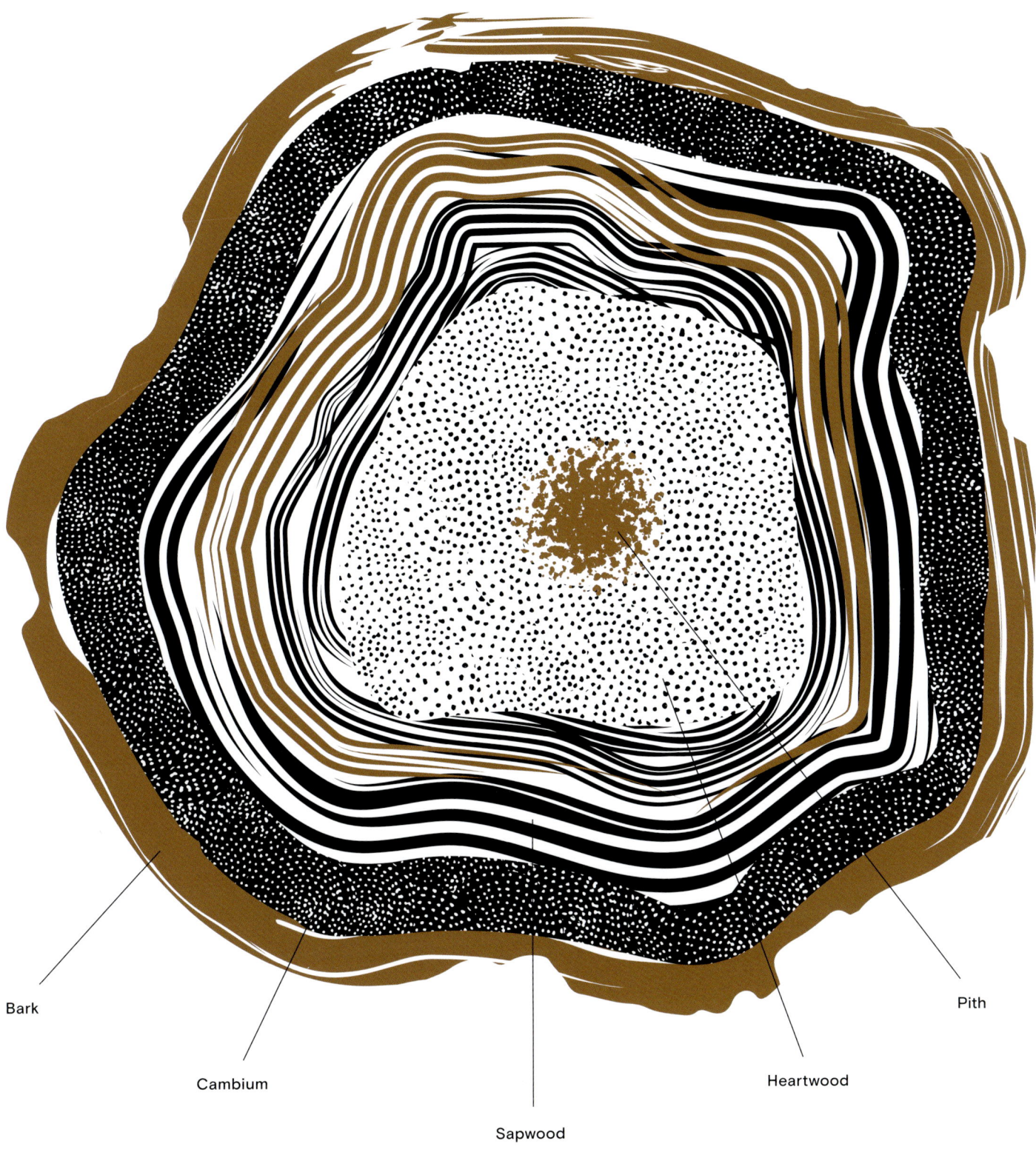

Bark

Cambium

Sapwood

Heartwood

Pith

Wood ⟶ Trunk Layers

Use of Timber through the Ages

Wood has been the material of choice for most of humankind's technological advances, providing the means for heat, shelter, transport, weapons, tools and homeware over the millennia. It is probably fair to say that without trees, we may not have survived as a species.

LES PREMIERS CANOTS DE LA SEINE

400,000 years ago
Wooden shelters

8000 BCE
Pesse canoe

1–2 million years ago
Firewood

40,000 years ago
Wooden rafts

4500–3300 BCE
Wooden wheels

RADEAUX DE L'ÂGE DE PIERRE

2000 BCE
Spoked wheels

1765
HMS *Victory*

2021
3D-printed homeware

753 BCE–476 CE
Canals, waterpipes, foundations

19th century
Wooden wagons

Wood Density

The 'Janka hardness test' was developed by an Austrian-born American researcher named Gabriel Janka (1864–1932) as a means to quantify the hardness of wood. It measures the force needed to embed a small steel ball into a wood sample.

↙ **Softwoods**
Softwoods, such as spruce (*Picea*) are fast-growing, reaching harvestable age in 30–40 years. For this reason, they are frequently planted in rotation and are usually derived from sustainable sources.

↓ **The timber trade**
Timber processing, at its most basic, involves cutting the tree trunk into flat planks. At one time this was a laborious, skilled manual job. Today, it is achieved with laser-guided electric band saws and round saws.

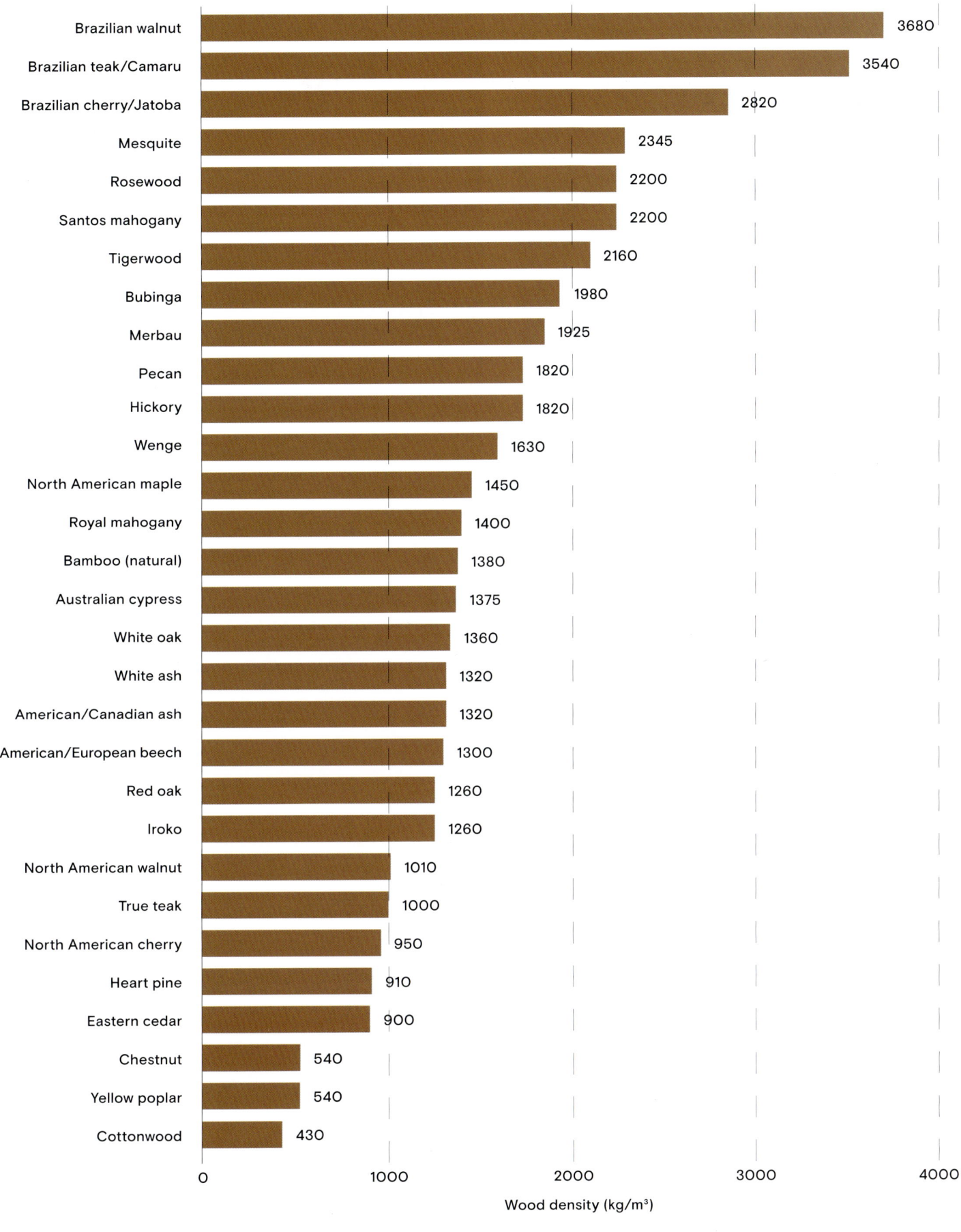

Wood density (kg/m³)

Colour

The colours of wood encompass a range of
rich hues, from black to brown to red to yellow.
The colours are enhanced by the texture
of the wood and its patterned veneer.

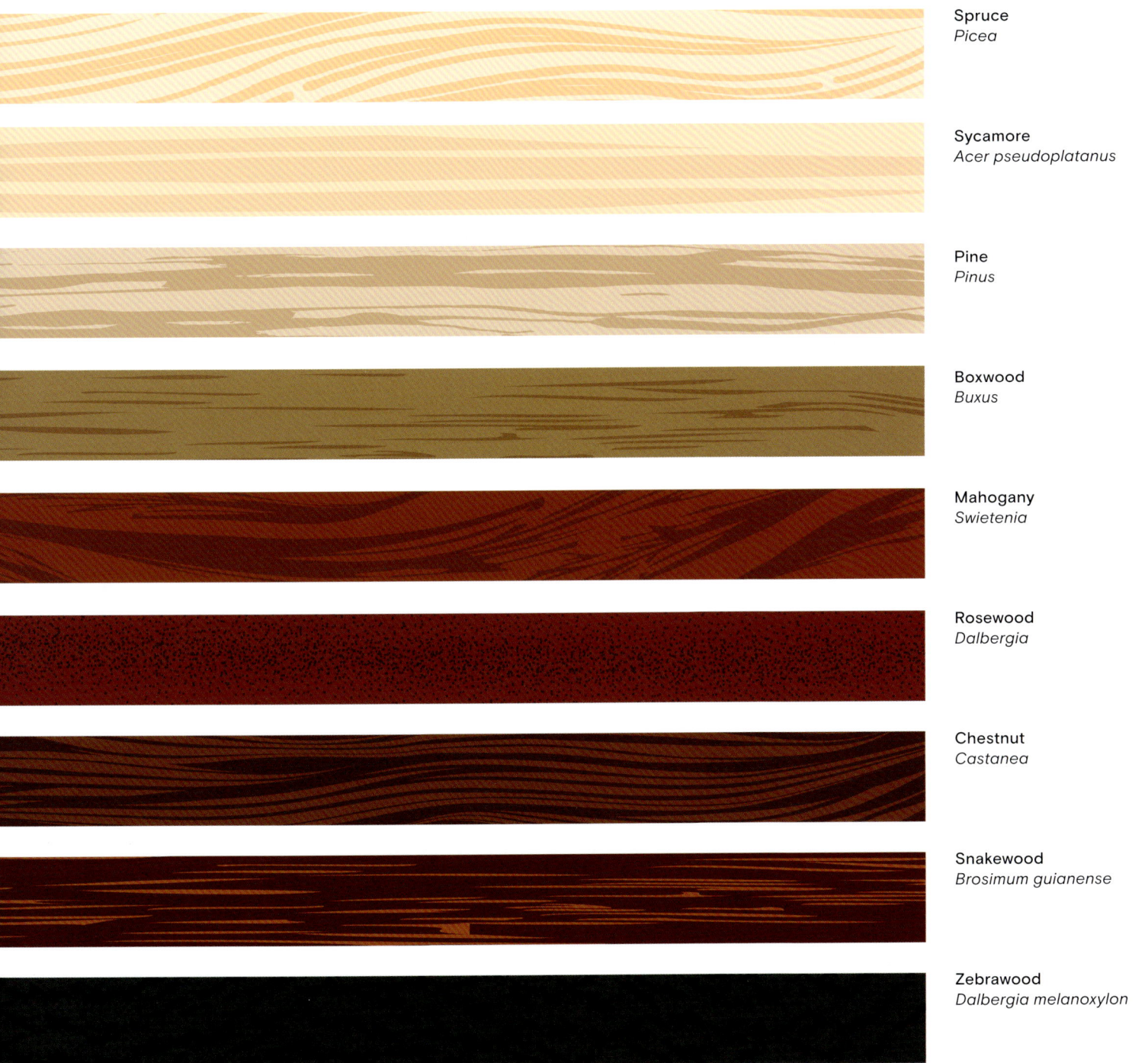

Spruce
Picea

Sycamore
Acer pseudoplatanus

Pine
Pinus

Boxwood
Buxus

Mahogany
Swietenia

Rosewood
Dalbergia

Chestnut
Castanea

Snakewood
Brosimum guianense

Zebrawood
Dalbergia melanoxylon

Types of Timber

Trees can be categorized as being either hardwoods or softwoods. Hardwoods are derived from deciduous, broadleaf trees like oak, maple, teak and mahogany, while softwoods come from evergreen conifers, such as pine, spruce and fir.

→ **Cherry tree**
Prunus serotina
Cherry-tree timber is a warm red colour that darkens with age. It is a choice material for crafting furniture and homeware, partly for its rich colour and also because it is fine-grained and easy to work.

→ **Caroline Place, London**
Amin Taha + Groupwork, 2012–17
This brick terraced house in London's Bayswater provided a blank canvas for a range of floor-to-ceiling utility spaces lined in cherry wood. Several of these are moveable and can be used to partition the rooms in different ways.

Oak

Teak

Maple

Mahogany

Hardwoods

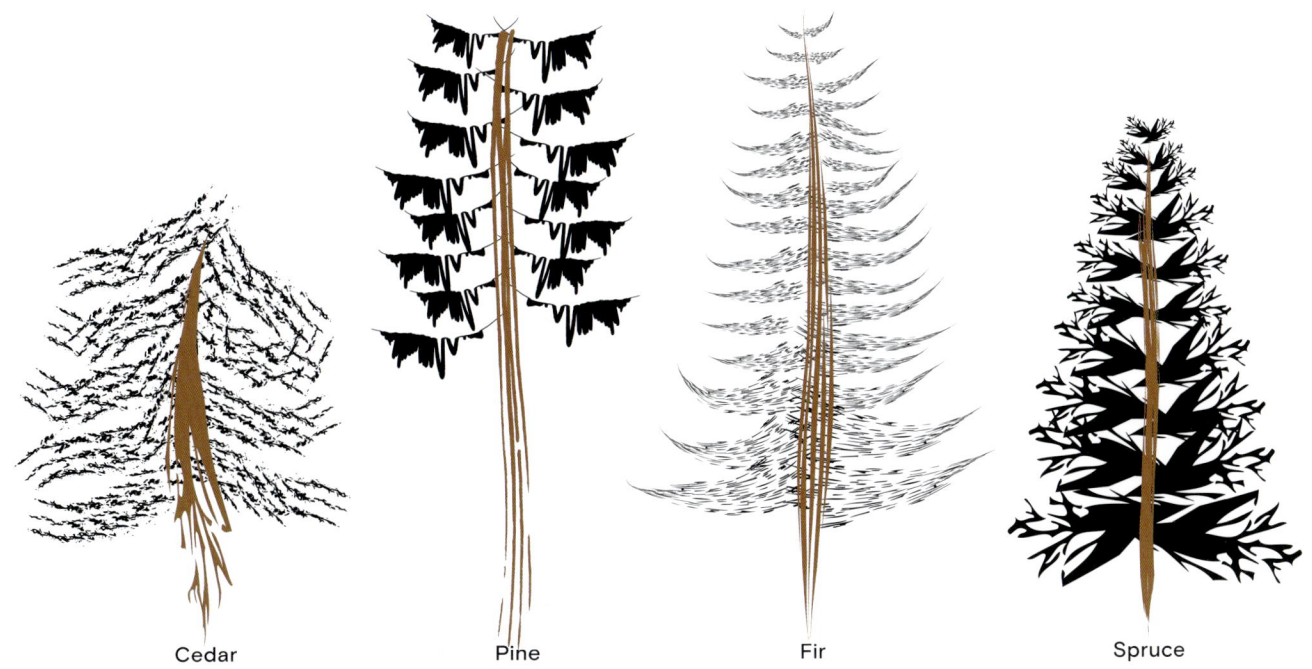

Cedar   Pine   Fir   Spruce

Softwoods

Architecture

As populations leave the countryside and migrate to urban centres, megacities of more than 10 million people are growing in number. The UN estimates that there are currently more than thirty megacities worldwide. For city planners and architects who understand that people need access to green spaces, this creates both challenges and opportunities.

1000 Trees Development, Shanghai
Heatherwick Studio, 2021
'The project has not only been conceived as a building, but as a piece of topography. Taking the form of two tree-covered mountains, "1000 Trees" is populated by hundreds of structural columns that emerge as large planters – each holding a cluster of trees.'

The Mopane Tree

If you fly over the south-central African bush a few weeks after a fire, you may see the spectral white shapes of trees etched into the landscape. These are the ghosts of central Africa's famous mopane trees (pronounced 'mopani'), the wood of which is so dense that if a tree catches alight, it will burn slowly like a cigarette, over many weeks, depositing the perfect outline of the living tree sculpted in ash. Mopane (*Colophospermum mopane*) usually grows almost as a monoculture (just the one species of tree in a given area), dominating woodlands with very few competing species. On deeper soils, mopane trees grow straight and tall, creating a cathedral-like effect, their trunks like pillars and their canopy forming a roof. In the autumn, mopane woodlands are similar in appearance to European beech forests, the trees' butterfly-shaped leaves turning every shade of yellow, orange and gold.

> The wood [...] is so dense that if a tree catches alight, it will burn slowly like a cigarette.

For many rural southern African communities that rely on firewood, mopane wood is like a low-grade coal. The wood is dense enough to sink in water – in fact, it is used as underwater décor in tropical fish tanks – and it burns slowly and steadily with very high calorific value, making it the perfect fuel for fires on which to cook and keep the house warm on a winter's night. When it burns, mopane wood also gives off a pleasant, aromatic smoke, which keeps mosquitoes at bay. Some years ago, my colleagues at Kew received some overseas development funding to assess the various firewood species used in Zimbabwe. After three years, and extensive experimentation, they concluded that mopane was the best firewood available. It occurred to me at the time that they could have saved a lot of time and expense by simply asking a local Zimbabwean.

↓ **Mopane**
Colophospermum mopane
Mopane (pronounced 'mopani'), from south-central Africa, is famous for the calorific value of its wood, which burns like a low-grade coal.

↘ **Burned trees**
These mopane trees in Moremi Game Reserve, Botswana, were burned in the dry season. The white shapes seen here are actually ash etched on the landscape.

Deforestation and Afforestation

According to Global Forest Watch, 411 million hectares (1,015 million acres) of forest have been cut down in the last twenty years. This represents an area the size of Greenland.

← Tropical hardwoods
Deforestation is highest in the Tropics where, critically, old-growth forests and hardwood timbers are not being replenished through planting.

↓ Tree nursery
Seedlings of only a few tree species are produced in nurseries. Less than one per cent of the world's tree diversity is grown by foresters.

Design and Technology

Yves Béhar's 3D-print Forust homeware range is made entirely from reclaimed wood waste. The process involves mixing sawdust with a natural tree-sap binder, which is then 3D-printed into complex, swirling geometries. The collection was created for an additive manufacturing company called Forust, who note that the approach is the 'first of its kind' for rematerializing offcuts from the wood and paper industry.

← Forust Vine Collection: 3D-print Forust homeware from reclaimed wood waste
Yves Béhar, 2021
Called Vine, the range includes a vessel, bowl, basket and tray, all 3D-printed from a sawdust composite.

↘ The designs replicate different wood grains and achieve a strength and durability that is apparently comparable to conventional timber.

'I see this material as a fantastic circular design opportunity to build unique products with additive technology, rather than subtractive; this will result in fewer trees needing to be cut and waste material being put to good use.'

— *Yves Béhar*

Flowers

✳ Flowers

The American poet and abolitionist John Greenleaf Whittier (1807–92) wrote: 'Give fools their gold, and knaves their power; let fortune's bubbles rise and fall; who sows a field, or trains a flower, or plants a tree, is more than all.'

This is a pleasing quote, and perhaps some consolation to foresters, farmers and horticulturists who can't compete with politicians or bankers for fame and fortune. Growing a tree that flowers might then be the closest a plants-man can come to fulfilment.

01

In John Whittier's day, wealthy landowners competed with each other to produce the newest, most spectacular rhododendrons, magnolias, peonies, hydrangeas or camellias. The 19th century was a golden age for horticulture and the plant collectors that fed the demand for new species – particularly flowering trees and shrubs. At the top of any estate-owner's wish list was *Rhododendron*, to the extent that 'rhododendromania' was a recognized phenomenon from the 1860s onwards.

There are over one thousand species of *Rhododendron*, including azaleas, and most of them produce large, showy, bell-like flowers in the northern spring. Their centre of diversity is temperate Asia, making them perfect for European and North American climates, and they are easily cultivated, particularly in acidic soils. The father of *Rhododendron* exploration and taxonomy was Sir Joseph Dalton Hooker (1817–1911), second Director of the Royal Botanic Gardens, Kew in London and close friend of Charles Darwin. Between 1848 and 1851, Hooker travelled extensively in South Asia, describing and collecting specimens of dozens of *Rhododendron* species. The seeds he brought back to Britain formed the basis of rhododendromania when they were hybridized with hardy species, and 'Himalayan dells' became fashionable features of Victorian estates. One of the best places to see an example of a Victorian 'rhododendron forest' today is at Cragside in Northumbria in the UK (now owned by the National Trust).

Plant hunting was not the sole preserve of scientists; nurserymen such as William and Thomas Lobb, E. H. Wilson, Francis Masson and Robert Fortune also brought thousands of plants into cultivation in the west. Perhaps the most adventurous of all was George Forrest (1873–1932), who spent twenty-eight years plant hunting in western China. His latter expeditions were funded by the Rhododendron Society, who were rewarded with over three hundred new species for cultivation.

The unluckiest plant hunter was undoubtedly David Douglas, a hardy Scot initially employed by Glasgow Botanic Garden and then by the Horticultural Society (now the Royal Horticultural Society), which was formed in 1804. Between 1823 and 1834, Douglas collected widely in North America – particularly the Pacific Northwest – and is credited with introducing more than two hundred new species into horticulture. He is best known for his conifer collections, including the Douglas fir (*Pseudotsuga menziesii*), but also introduced some wonderful flowering trees and shrubs, including the silk tassel bush (*Garrya elliptica*) and the flowering currant (*Ribes sanguinium*). On first disembarking at New York in August 1823, he was initially refused entry for being too scruffy – a stroke of misfortune among the many that would plague his plant-collecting days. Amongst these were a horse that bolted with him on its back (according to Douglas, the horse apparently only understood French, which he did not speak); having all of his belongings stolen while he was up a tree collecting seeds; and his boat nearly sinking in a storm. Encountering wild weather was a particular hazard, especially in the Pacific Northwest, but he seems to have struck up a rapport with the Native Americans, who often guided Douglas and got him out of trouble – though one tribe proclaimed him to be an evil spirit when they saw him drinking a fizzy health tonic, which they mistook for boiling water. Perhaps as a result of his various deprivations, his eyesight deteriorated, and eventually he lost his sight in one eye completely. This may have contributed to his demise, which came about in Hawai'i on 12 July 1834, when he stumbled into a pit holding one of the island's wild cattle. His trampled and gored body was found later that day by some locals.

> At the top of any estate-owner's wish list was *Rhododendron*, to the extent that 'rhododendromania' was a recognized phenomenon from the 1860s onwards.

Probably my favourite plant hunter of the Victorian era was Marianne North, who was neither a scientist nor a horticulturist, but an artist. Born in 1830 and eldest daughter to Frederick North, Member of Parliament for Hastings in Sussex, she grew up surrounded by high-profile literary and artistic types, including Edward Lear, Sir William Hooker, William Hunt and Lucie Austin, later to become Lady Duff Gordon. She was thirty-nine and unmarried when her father died and, having travelled extensively as his companion when he was alive, she inherited enough money to be of independent means for the rest of her life. Undaunted by the fact that unmarried Victorian ladies were not supposed to travel unaccompanied, she planned and undertook a series of epic journeys right across the world for the next sixteen years, painting the landscapes, flowers and trees she encountered in striking colour. In this age before colour photography, her beautiful oil paintings created a kind of virtual grand tour encompassing the floras of North America, South America, Europe, Asia, the Indian Ocean, Africa, the Pacific and Australasia. Her travelogue and memoirs, first published in 1893, three years after her death, were aptly entitled *Recollections of a Happy Life*, and her joy comes across clearly in her paintings and writing.

In 1879, in between trips to India and Australasia, Marianne North wrote to Sir Joseph Hooker, then Director of Kew Gardens, and offered to build a gallery at Kew to house her paintings. Hooker was delighted and accepted the proposal; Marianne was allowed to supervise the plans for the building and the arrangement of her pictures. Her gallery, housing some eight hundred paintings, opened in 1882 and can still be visited at Kew today. I have always considered it to be something of a hidden treasure because it is some distance from the main Victoria and Elizabeth gates, and from the outside the gallery looks unprepossessing compared to Kew's great glasshouses and other heritage buildings. Go inside, however, and you are rewarded with wall-to-wall colour and a unique, extraordinary rendition of the world's flora. Her tree pictures are particularly fine, and include Indian *Rhododendrons*, a Brazilian coral tree (*Erythrina*) and a Chilean southern beech (*Nothofagus obliqua*). When guiding guests around Kew, we frequently set a challenge for visitors to the gallery: to find the only self-portrait of Marianne North known to exist. Something of a trick question, the portrait hangs on the wall opposite the front door to the left, and shows a tiny figure dressed head to toe in black being carried in a sedan chair during her travels in Asia.

In general, bees respond to the blues, purples and pinks. Moths, on the other hand, prefer white, cream and lighter hues, while wasps, ants and flies are attracted to yellows. Beetles and butterflies prefer reds and blues.

02 – Common juniper
Juniperus communis
Juniper trees are dioecious,
meaning that male and
female flowers grow on
different trees. Male and
female juniper flowers are
small and inconspicuous,
a clue that this species is
pollinated by wind. Female
flowers develop into juniper
berries, used to flavour gin.

02

Flower Shape, Colour and Scent

As with all blossoming plants, the main objec-
tive of a tree's flower is to be fertilized. In some
species, male and female flowers grow on separate
trees, which are termed 'dioecious'. The advantage
of this arrangement is that trees cannot fertilize
themselves, thereby avoiding inbreeding. The
disadvantage is that a tree of the opposite sex must
be within reach of pollination. Holly (*Ilex aquifolium*)
is an example of a dioecious tree species, and it's
only the female trees that produce the berries
we use as decorations at Christmas time. Another
dioecious species is southern Africa's marula tree
(*Sclerocarya birrea*), the fruit of which is much
loved by elephants [pg.230]. If your house is built in
the shade of a female marula tree, then come May,
watch out – you can expect some large visitors when
the tree is in fruit. Other tree species produce male
and female flowers in separate structures on the
same tree, but usually at different times, helping
to minimize self-fertilization; these species are
called 'monoecious'. Birch (*Betula*), hickory (*Carya*),
chestnut (*Castanea*), hazel (*Corylus*), beech (*Fagus*)
and walnut (*Juglans*) are examples of monecious
trees. Finally, many species produce hermaphrodite
flowers that have both female (ovary, style, stigma)
and male (stamen) parts [pg.192].

The shape (or 'morphology'), colour and scent of a tree flower gives a strong indication of how the tree is pollinated [pg.190]. Species with small, inconspicuous flowers, without showy, colourful petals, are likely to be wind-pollinated, and often produce their flowers in spikes or catkins that hang down from the branches to help the pollen catch in the breeze. Examples of wind-pollinated trees include birch (*Betula*), hazel (*Corylus*) and oak (*Quercus*). In contrast, trees that produce large, colourful flowers are likely to be pollinated by birds, insects or mammals. Most animal-pollinated plants are pollinated by insects and, although colour is an important cue (together with shape and odour) for pollinating insects, it isn't necessarily the main draw. In general, bees respond to the blues, purples and pinks. Moths, on the other hand, prefer white, cream and lighter hues, while wasps, ants and flies are attracted to yellows. Beetles and butterflies favour reds and blues. Flower colour is also an indication of fertility, with bright colours fading as the flower ages. However, some trees produce flowers that actually change colour, attracting different pollinators over time. For example, the Chinese honeysuckle (*Quisqualis indica*) produces flowers that are initially white before turning pink and then red. The white flowers have been shown to attract moth pollinators at night, the pink flowers bees and the red flowers butterflies. Through this mechanism, the plant attracts a range of pollinators over several weeks, and is not solely reliant on one pollinator group.

According to Bat Conservation International, bats pollinate more than 530 species of plant, including tropical fruit trees such as mango, guava, durian and banana.

03

Where a plant does rely on a single, specialized pollinator, there can be reproductory advantages. This represents a form of 'co-evolved mutualism' – or 'symbiosis' – with other species, an idea that is explored more thoroughly in Chapter 8 [pg.238]. An extreme example of this is seen in fig trees (*Ficus*), where the inside-out arrangement of the flowers is the result of co-evolution with certain kinds of wasp.

03 – Spiderhunter
Arachnothera zeylonica
The spiderhunter shown in this print is a type of sunbird. Along with hummingbirds, sunbirds are important pollinators of tree flowers.

04 – Fig tree
Ficus
Inside every fig is a hollow chamber called a syconium that holds hundreds of tiny flowers.

04

Next time you eat a fig, break it open and examine it carefully. Inside the fruit, you will find hundreds of tiny seeds – in fact, the whole fruit is an inside-out bunch of miniscule flowers. The hollow chamber inside the fig is called a 'syconium', and the flowers project into it. Specialized wasps enter the syconium through a small opening, where they fertilize the flowers and lay their eggs. The syconium then acts as a nursery for the baby wasps. When the eggs hatch, they produce wasps of both sexes; the males mate with the females, and then create an opening for the females to escape. The males then die, their job done, and the flowers mature into fruits and the cycle repeats itself. There are around 750 species of fig tree [pg.186] and each has its own pollinating wasp species.

Flower scents are primarily produced by plants to attract insect pollinators, with bees, wasps, moths and butterflies all attracted by the smell of sweet nectar, and those pollinated by beetles attracted by strong musty, spicy or fruity odours. In Zambia's Luangwa Valley, home to perhaps the greatest concentration of wildlife in Africa, tourists travelling back to their lodges after evening game drives in June or July will invariably encounter the all-pervading scent of the potato bush (*Phyllanthus reticulatus*). The flowers of this pan-tropical small tree are almost invisible to the naked eye but after nightfall they emit an extremely powerful smell akin to baked potatoes. So elusive is the source of the scent that African folklore posits it as the smell of snakes – which can be sensed but never seen. In fact, the scent attracts several species of pollinating moth in the genus *Epicephala*.

In contrast to insects, birds don't have a well-developed sense of smell and so scent is not generally a factor in bird pollination. Bats, on the other hand, are attracted by musky, sulphurous scents emanating from pale flowers that can be seen at night. According to Bat Conservation International, bats pollinate more than 530 species of plant, including tropical fruit trees such as mango, guava, durian and banana. Declines in bat populations therefore have a negative impact on all these crops, which can also hamper local economies.

For trees with a specialist pollinator, it is essential that the timing of flowering ('phenology') coincides with the right point in the pollinator's life cycle: if insect-pollinated trees flower when there are no insects around, they won't be fertilized. Sadly, the activities of humans have sometimes disrupted this delicate synchrony. The use of 'neo-nicotinoid' insecticide seed coatings for crops, such as oilseed rape (canola) or in sprays used on fruit trees, has been shown to increase mortality in honeybees by impairing their homing ability and to reduce the reproductive success of bumblebees and solitary bees. In 2018, the European Union banned the use of the main three neonicotinoids outside, mainly because of concerns about plummeting bee populations which, in turn, were having a major impact on the soft-fruit industry. As a result of bee shortages, many fruit growers have resorted to importing bees from countries where their populations are not affected. Unfortunately, this can also lead to the import of bee parasites (e.g. mites) and diseases.

Global warming has also had an impact on phenology. In a study carried out in 2006, which analysed 125,000 phenology records for 542 European plant species, researchers found that on average leaf and flower appearance advanced by 2.5 days per decade between 1971 and 2000.

For generalist tree species that are pollinated by the wind or multiple insect species this is not a problem, but for species that rely on a single pollinator, flowering early might mean losing the services of their pollinator. The Japan Meteorological Agency has monitored the flowering of several cherry-tree (*Prunus*) species and the appearance date of the pollinating butterfly *Pieris rapae* in spring at Nagano, Japan, since 1953. Flowering has tended to occur earlier over the last three decades, whereas the appearance of the butterfly has been delayed.

The cultural importance of cherry-tree blossom in Japan is discussed on pg.198, but another example of how flowers have shaped human discourse can be seen in the case of the UK's Glastonbury thorn (*Crataegus monogyna* 'biflora'). Unlike the wild hawthorn, this particular cultivar flowers twice a year – once in spring (as is usual) and again in winter at Christmas time when there are no pollinating insects to be seen. This peculiar trait has strengthened the myth surrounding the Glastonbury thorn: that shortly after the death of Jesus, Joseph of Arimathea came bearing the Holy Grail and a staff, which he thrust into the ground and from which the thorn tree grew. The origin of the legend can be traced to the Middle Ages and was likely a ruse to bring pilgrims to the abbey at Glastonbury. The original tree was destroyed during the English Civil War in 1647, probably by a Parliamentary soldier because of its associations with Catholicism and Christmas. The story goes that before it was destroyed, secret cuttings were made of the holy tree – and so its descendants still survive in St John's Churchyard and maintain the 'miraculous' habit of flowering at Christmas time.

Figs

The fruit of a fig has a green skin that deepens
to a rich purple as it ripens. Inside, the fig is
essentially an inside-out bunch of hundreds of
minute flowers. These are fertilized by specialized
fig wasps that enter through a tiny opening.

← Fig
Sarah Thompson-Engels,
2009

Oil sticks and pastels.

↓ Fig
Johann Jakob Haid,
1704-67

Coloured engraving.

Blossoms

Most blossom trees are stone-fruit trees (*Prunus*), though the name can also be applied to plants of a similar appearance and flowering pattern. The majority of these species are pollinated by insects, for whom colour and scent are important cues.

Brazilian coral tree
Erythrina falcata

Sweet cherry
Prunus avium

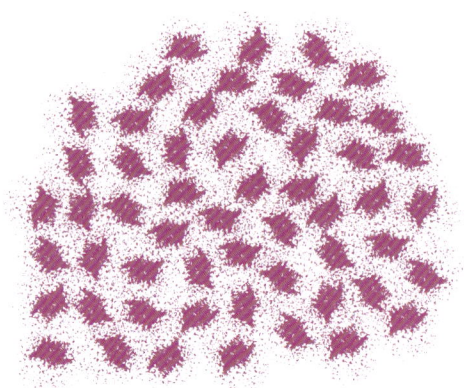

Rhododendron
Rhododendron ponticum

Yellow azalea
Rhododendron luteum

Bellflower cherry
Prunus campanulata

Rhododendron hookeri
Rhododendron hookeri

Golden camellia
Camelia chrysantha

Weeping higan cherry
Prunus subhirtella var. pendula

Oshima cherry
Prunus speciosa

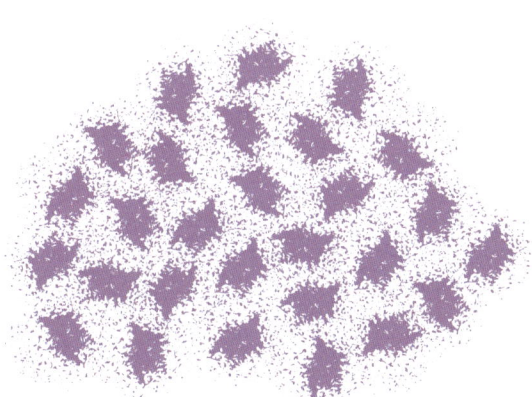

Purple rhododendron
Rhododendron impeditum

Almond
Prunus dulcis

Pollination Methods

Trees can be pollinated in several different ways. How a tree's flowers are shaped, coloured and scented all give clues to its chosen method of pollination. Small and muted flowers suggest a species is wind-pollinated, while big, bright blossoms are designed to attract animals – birds, insects or mammals – that transport the pollen.

01

02

03

04

01 – Bird pollinated
Bromeliad, Bromeliaceae
The hummingbird is an important pollinator of bromeliads.

02 – Wind pollinated
Pine, Pinus
Pines produce huge amounts of pollen, which is carried on the wind.

03 – Bee pollinated
Apple tree, Malus
Bees are essential to commercial fruit growers.

04 – Moth pollinated
Yucca flower, Yucca torreyi
The yucca flower is exclusively pollinated by the yucca moth (*Tegeticula antithetica*).

05 – Bat pollinated
Parry's century plant, Agave parryi
Bat pollination is economically important for tequila production.

06 – Gecko pollinated
Bromeliad, Bromeliaceae
Gecko lizards pollinate many tropical plants, including bromeliads and screwpines.

07 – Bee pollinated
Mimosa, Acacia dealbata
Bees pollinate many tree species, including the mimosa tree, which has bright yellow flowers.

08 – Butterfly pollinated
Buddleia, Buddleja davidii
Favoured by butterflies, the buddleia tree is commonly known as the butterfly bush.

05

06

07

08

Agave parryi, Bromeliaceae, Acacia dealbata, Buddleia

Anatomy

The anatomy of a flower includes the male parts (the anthers and filaments) that produce pollen and the female parts (the stigma, style and ovary) that receive the pollen.

Foliage, flowers and fruit of a common Indian forest tree (*Phanera variegata*)
Marianne North, 1878
Oil on board.

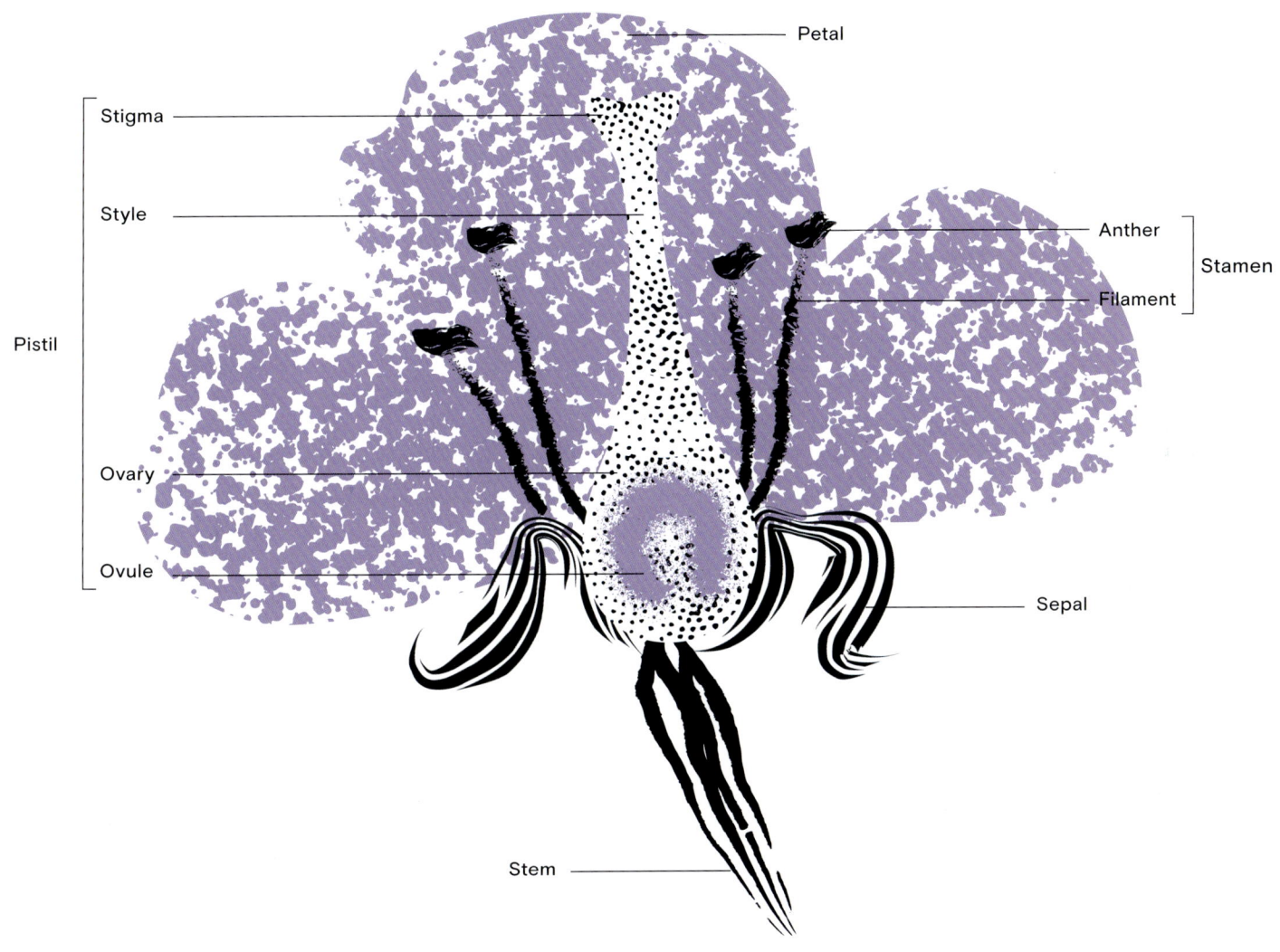

Petal

Stigma

Style

Anther

Stamen

Filament

Pistil

Ovary

Ovule

Sepal

Stem

Origin of Different Flowering Tree Species

Flowering plants ('angiosperms') evolved 130 million years ago and diversified enormously during the Cretaceous period. Angiosperms are the most diverse group of land plants, with around 370,000 known species. Species diversification is highest in isolated geographic locations like the island of Madagascar, where more than 90 per cent of its 14,000 plant species occur nowhere else.

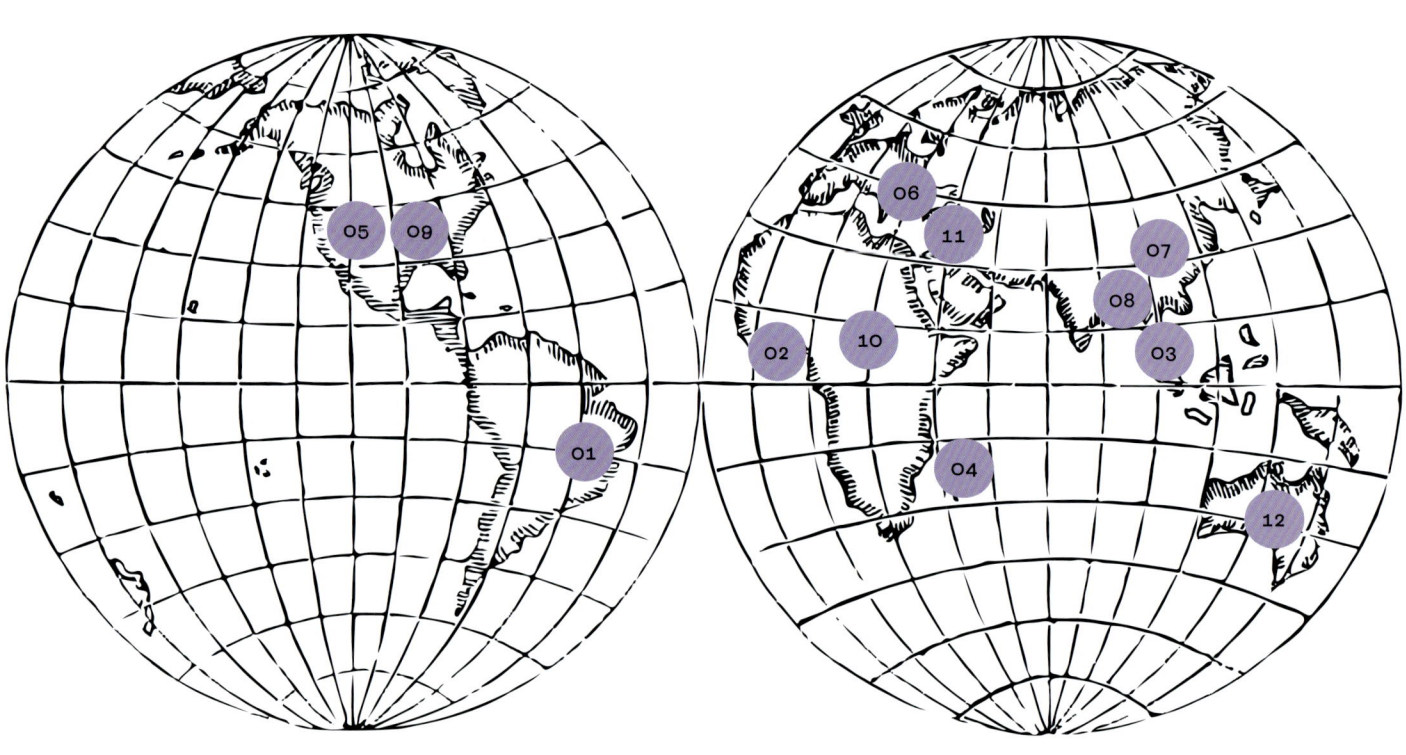

01 – Jacaranda mimosifolia, *Brazil*

02 – Spathodea campanulata, *Ghana*

03 – Cassia siamea, *Thailand*

04 – Delonix regia, *Madagascar*

05 – Magnolia grandiflora, *USA*

06 – Aesculus hippocastanum, *Balkans*

07 – Davidia involucrata, *China*

08 – Rhododendron hookeri, *Bhutan*

09 – Franklinia alatamaha, *USA*

10 – Kigelia africana, *Central Africa*

11 – Rhododendron ponticum, *Turkey*

12 – Corymbia ficifolia, *Australia*

Japan's Cherry Blossom Forecast

The blossoming of cherry trees is of huge cultural significance in Japan. The appearance of the 'sakura' cherry blossoms, heralding the coming of spring, is forecast like the weather.

Cherry Blossoms, Tama River Embankment
Ando Hiroshige, 1797–1858
From 'One Hundred Views of Famous Places of Edo'. Japanese print.

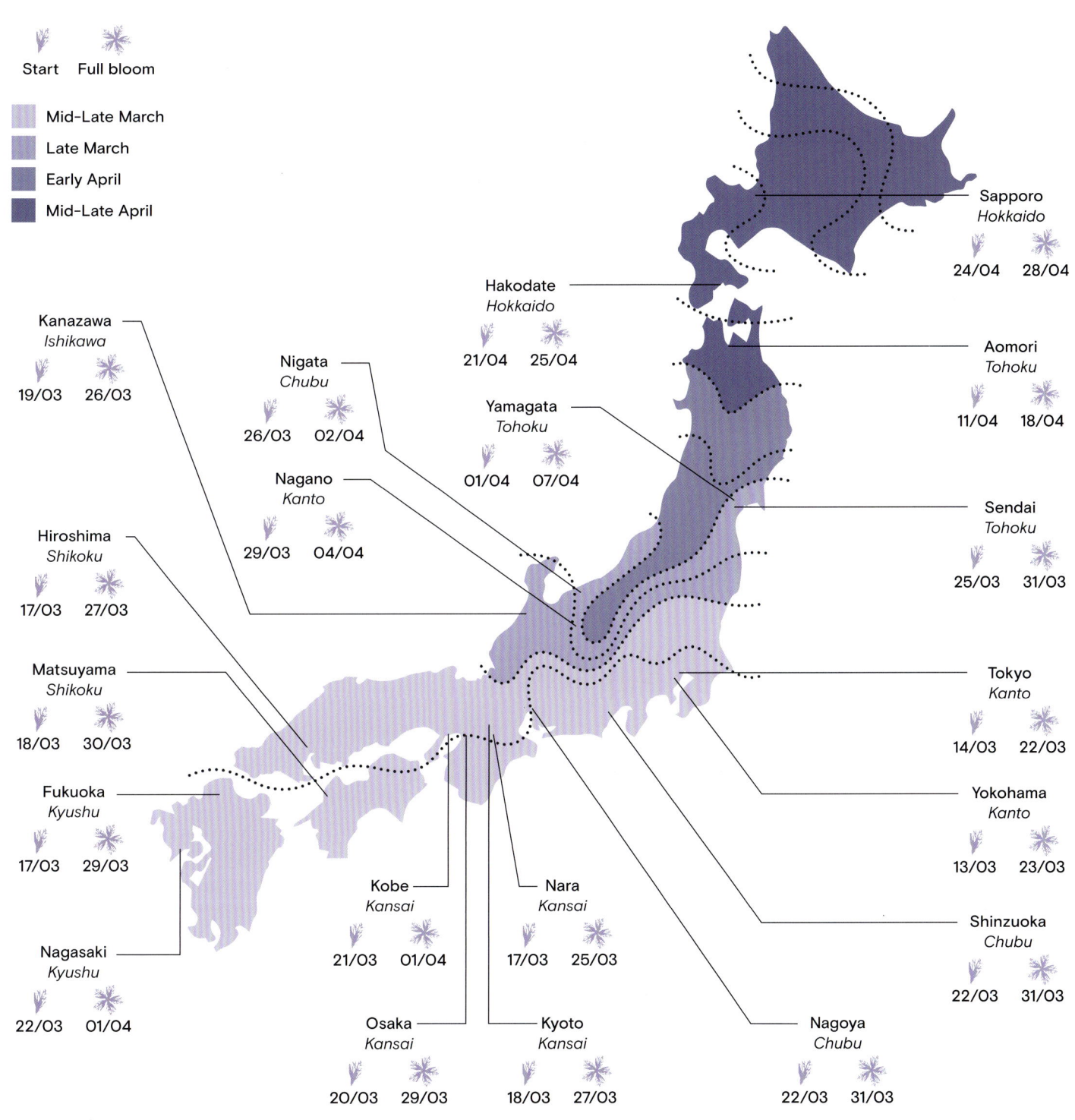

Start **Full bloom**

Mid-Late March
Late March
Early April
Mid-Late April

Sapporo
Hokkaido
24/04 28/04

Hakodate
Hokkaido
21/04 25/04

Aomori
Tohoku
11/04 18/04

Kanazawa
Ishikawa
19/03 26/03

Nigata
Chubu
26/03 02/04

Yamagata
Tohoku
01/04 07/04

Sendai
Tohoku
25/03 31/03

Nagano
Kanto
29/03 04/04

Hiroshima
Shikoku
17/03 27/03

Matsuyama
Shikoku
18/03 30/03

Tokyo
Kanto
14/03 22/03

Yokohama
Kanto
13/03 23/03

Fukuoka
Kyushu
17/03 29/03

Shinzuoka
Chubu
22/03 31/03

Nagasaki
Kyushu
22/03 01/04

Kobe
Kansai
21/03 01/04

Nara
Kansai
17/03 25/03

Osaka
Kansai
20/03 29/03

Kyoto
Kansai
18/03 27/03

Nagoya
Chubu
22/03 31/03

Flower-viewing in Japan

Japan's tradition of 'hanami' or 'flower-viewing' goes back millennia – at least to the Nara period of the 8th century, and probably much further than that. Festivals that celebrate the transient flowering of fruit trees in Japan mainly centre around the cherry (sakura) but may also include plum trees (ume). In the past, the flowering of the sakura signalled the start of the rice-planting season, and was also used to divine the harvest, with offerings being made to the Shinto spirits or 'kami' inhabiting the trees. In fact, some individual trees of great antiquity are found on Buddhist and Shinto temple sites, and are themselves venerated to this day.

In 2012, when I was working at Kew's Millennium Seed Bank (MSB), we were visited by the Japanese Ambassador and a delegation from Fukushima, which had suffered a devastating tsunami the year before. They brought with them seeds from the 'Miharu Takizakura' cherry tree, located near their town; the tree was around one thousand years old and had survived the tsunami. The delegation brought the seeds to the MSB in gratitude for British aid during the disaster but also for safekeeping in the vaults of the Seed Bank [pg. 42].

This gift was deeply symbolic, as in Japan the transient, ephemeral splendour of the country's blossom trees is a metaphor for life itself. The fact that the 'Miharu Takizakura' cherry tree survived the tsunami was a reminder that the beauty of nature frequently transcends human life and suffering.

The 'Miharu Takizakura' tree is a weeping higan cherry (*Prunus subhirtella var. pendula* 'Itosakura'), the oldest cherry cultivar in Japan, dating from at least the Heian period (794–1185). These specimens have the longest lifespan among Japanese cherry cultivars and are easy to grow into large trees. For this reason, there are many large and long-lived examples of this species in Japan, and they are often regarded as sacred for their association with Shinto and Buddhist temples. In the Kamakura period (1185–1333), the Oshima cherry (*Prunus speciosa*), originating from Izu Oshima Island, became popular in cultivation. Then, in the Muromachi period (1336–1573), the 'sato-kazura' group of hybrids, based on the Oshima cherry, began to appear.

Finally, in the Edo period (1600–1868), a large number of double-flowered cherry cultivars were produced – books from that period record more than two hundred varieties of cherry-blossom cultivars.

Today, hanami is as popular as ever, and is an opportunity to cele-brate the coming of spring. The 'cherry-blossom front' is predicted in Japan like a weather forecast: it is followed on TV and radio programmes as it advances from the south,

> The 'cherry-blossom front' is predicted in Japan like a weather forecast: it is followed on TV and radio programmes as it advances from the south, starting on the sub-tropical islands of Okinawa and moving northwards throughout March and April.

starting on the sub-tropical islands of Okinawa and moving northwards throughout March and April. Cherry-blossom parties are usually held outside in public parks, and night-blossom festivals ('yozakura') have become popular too, accompanied by the lighting of paper lanterns. Like most celebrations, hanami is an occasion to eat and drink, with rice wine (sake) being the tipple of choice, but prayers and singing are also common traditions. As well as being important domestically, Japanese blossom gardens represent a major cultural export for Japan. Nearly every city in the world has a Japanese garden and, although many of these are derivatives, conceptually they follow Japanese naturalistic aesthetics, which eschew straight lines, mimic wider landscapes and combine seasonal colour with artistic hard landscaping and water features.

Hanami

In Japan, 'hanami' – translated as 'flower-viewing' – is centred around the country's abundant cherry trees (sakura) – and, less commonly, plum (ume) trees. This tradition dates back at least to the Nara period around the 8th century.

Chiyoda Great Interior Flower Viewing
Toyohara Chikanobu, 1894

A tryptych in the ukiyo-e style depicting a flower-viewing party in the harem of Edo Castle at Chiyoda.

Fashion

Christian Dior's Spring–Summer 2020 collection fashion show featured an 'inclusive garden' – a set that included a temporary grove of 164 trees. This represented a collaboration with Coloco, a collective of botanists, gardeners and urban landscapers whose philosophy is to 'facilitate active exchange between citizens and nature' through communal gardens. The trees were later donated to sustainability projects in the Paris region, bringing new greenery into the city's heart.

Christian Dior runway
Paris Fashion Week, 2019

For the company's spring/summer 2020 collection, shown during Paris Fashion Week, creative director Maria Grazia Chiuri was inspired by photographs of Christian Dior's sister Catherine showing her surrounded by flowers in her garden.

'Christian Dior loved gardens. Flowers are part of the heritage of Dior. But when I see flowers and gardens now, I don't see just joy, or the beauty of nature. I see all the concerns about our future, and the need to take action [...]. So I was asking myself, how can we celebrate nature in a meaningful way?'

— *Maria Grazia Chiuri,* Creative Director at Dior

Design and Technology

All the pieces in Dutch designer Matilde Boelhouwer's series of artificial flowers – called 'Insectology: Food for Buzz' – turn rain into sugar water. Boelhouwer's five flowers were generated using screen-printed polyester, and each serves as an emergency urban food source for a different one of the 'big five' pollinators – bees, bumblebees, hoverflies, butterflies and moths.

↓ Insectology: Food for Buzz
Matilde Boelhouwer, 2019
There are five flowers in the series, each featuring a different colour scheme and design. 'I did research into why insects are attracted to specific flowers and there is more than one reason: its colour, its shape, and its smell,' said Boelhouwer.

→ Butterfly design
As butterflies tend to have long tongues (2–5 cm/0.8– 2 in.), they need more time to feed. They therefore typically choose flowers with a larger bottom petal to give them more stability to rest – replicated in this artificial-flower design.

Fruits

Fruits

Fruits come in all different shapes and sizes, colours and textures. Designed to be eaten by creatures great and small, these fleshy structures contain the tree's seeds. The majority of fruits that are consumed by humans are only eaten locally – meaning many varieties are widely unknown.

01

01 – Sausage tree
Kigelia africana
The fruits of this African tree can weigh up to 22 kg (45 lb). Although not edible to humans, they are used for making traditional beer.

Sir Isaac Newton famously discovered the law of gravity when an apple fell on his head. Had Newton been sitting under a different kind of tree, he may never have theorized again. The 'Sausage Tree' (*Kigelia africana*), which is found along southern African rivers, for example, produces fruits that weigh up to 22 kg (45 lb). The purpose of fruits is to enable the dispersal of seeds, so there is always a good reason for fruit being shaped a certain way. In the case of the fruit of the sausage tree, its large size is probably to make it more attractive to the big beasts that dine on it – hippos, rhinos and elephants, all of which are champion seed dispersers in the African bush.

As with all things botanical, botanists have created a huge lexicon of names for fruits based on their function or 'morphology' – berry, 'cypsela', 'drupe', 'hesperidia', 'pepo', 'pod', 'pome', 'samara' and 'syncarpium', to name a few. In this book, we confine ourselves to a few types of fruit that are both found on trees and have entered into common parlance. A drupe is a fruit with a hard coated 'stone', typically surrounded by a fleshy pulp and outer skin layer. Plums, cherries, apricots, mangos and peaches are all drupes. Berries, on the other hand, have more than one seed embedded in a fleshy pulp, examples being persimmon, blackcurrants, grapes and elderberries. Citrus fruits are a kind of berry but are segmented, and are called hesperidia – admittedly not a word you hear every day. Pomes take us back into more familiar territory thanks to the French word 'pomme', meaning 'apple'. Apples, pears, quince and rowan are all pomes – fruits with relatively hard flesh that surrounds a core containing seeds. Other common kinds of tree fruits are pods, which may or may not split open to release seeds, and samaras, which are winged to allow wind dispersal. The sausage tree mentioned above is something of an anomaly, most closely resembling a pepo – fruits that have multiple seeds throughout the flesh or grouped together in the centre like a melon or a cucumber. In fact, sausage-tree fruits are modified pods that, unlike most of the pods in their family, *Bignoniaceae*, do not split open.

Why Are Fruits Important?

To the tree, fruits are means of dispersing their seeds, but in doing so, they provide food for a vast array of species from micro-organisms up to the largest mammals, including humans [pg.226]. In natural ecosystems, fruit trees are the factories that build habitats. They attract animals and birds that feed on their fruit, creatures that in turn deposit the seeds of other fruits they have eaten – leading to a proliferation and diversity of flora in the vicinity of fruit trees. As humans attempt to restore complex habitats, fruit trees are crucial to the so-called 'framework species' methodology. This approach entails the planting of 'keystone' pioneer and fruiting species first, in order to kickstart succession and help establish more species. For this sort of habitat building to be successful, however, you need patches of natural forest nearby from where birds and animals can transport seeds.

A typical keystone species is the fig (*Ficus*, pg.186). More than 1,200 different species of bird and mammal forage on figs, and they feed more species of bird and insect than any other fruit. Many of the world's 750 fig species also produce at least two crops of figs each year, and the largest fig trees produce a fruit crop measured in tonnes rather than kilogrammes. Figs are undoubtedly fruiting superstars, but even the humblest flowering tree or conifer produces fruits that are eaten by something. Multiply all those beneficiary species by the roughly sixty thousand tree species in the world, and you can start to see how important tree fruits are at the base of the food chain.

> More than 1,200 different species of birds and mammal forage on figs, and they feed more species of bird and insect than any other fruit.

Trees as Food

When it comes to eating the fruits or seeds of plants, humans are pretty conservative. Around 50 per cent of our plant-based calorie intake comes from eating the seeds of just three grass species – wheat, maize and rice – while 80 per cent is accrued from the seeds or tubers of twelve plant species: wheat, maize, rice, barley, oats, millet, sorghum, finger millet, potato, sweet potato, yam and taro – none of which are tree species. In total, it is thought that humans eat no more than 200 species of plant, and yet some experts estimate that we could eat 300,000 of the 400,000 total.

Bananas, apples, oranges, mangoes, apricots, tangerines and peaches are the most popular commercially traded fruits (in order) that grow on trees [pg.222], with global banana production topping 115 million metric tonnes (113 million imperial tons) in 2019. However, these figures only relate to formal markets. Much more fruit, from a greater variety of species, is eaten or traded informally by subsistence farmers

> In total, it is thought that humans eat no more than 200 species of plant, and yet some experts estimate that we could eat 300,000 of the 400,000 total.

and rural peoples. For example, the musuku tree (*Uapaca kirkiana*) of south-central Africa and the closely related tapia (*Uapaca bojeri*) of Madagascar produce sweet, plum-sized fruits that taste like stewed apricots, and are extremely important nutritionally in remote areas. Similarly, Africa's mobola plum (*Parinari curatellifolia*) produces a plum-like fruit relished across large swathes of the continent. In fact, so important is the mobola plum tree that Victorian explorer David Livingstone's heart was buried beneath one when he died at Chief Chitambo's village in 1873.

One of the better-known 'minor' fruit crops is breadfruit (*Artocarpus altilis*), which belongs to the same plant family as the ubiquitous fig. It is derived from the wild species *Artocarpus camansi* – native to New Guinea, the Maluku islands and the Philippines. From its home range, domesticated varieties of breadfruit were taken to the South Pacific by early Austronesian migrants, and then to the New World by Europeans. In 1787, the infamous British HMS *Bounty* [pg.218] was commissioned to transport breadfruit from Tahiti to the West Indies to see if it would grow there. HMS *Bounty* never arrived in the Caribbean; half of the crew, led by Acting Lieutenant Fletcher Christian, mutinied near Tonga and set Captain Bligh and his officers adrift in the *Bounty*'s launch. In a remarkable feat of navigation, the 19 loyal officers and seamen sailed 3,500 nautical miles in an open boat 7 metres (23 feet) long to the Dutch settlement of Coupang (now Kupang) on the island of Timor. The mutineers, in the meantime, eventually made a home on the Pitcairn Islands, where their descendants still live. There were two botanists aboard HMS *Bounty*: David Nelson and William Brown. Nelson remained loyal to Bligh and made it to Coupang but unfortunately died a few days later after succumbing to fever. Tasmania's Mount Nelson is named in his honour. William Brown, meanwhile, was amongst the mutineers who travelled to the Pitcairn Islands, where he was killed by the Polynesian nobleman, Minarii, on 'Massacre Day' (20 September 1793). Ironically, although breadfruit was eventually taken to the West Indies, it never took hold as a crop there, the locals preferring plantains and other staples.

Closely related to the breadfruit is its spiky cousin, the jackfruit (*Artocarpus heterophyllus*), which is native to Malaysia, India and Sri Lanka. It is widely consumed in south and southeast Asia and is the national fruit of both Bangladesh and Sri Lanka. Its sweet flesh is likened to that of pineapple or banana; the boiled, baked or roasted seeds are also edible, and taste similar to the Brazil nut. Like the breadfruit, the jackfruit grows to an enormous size, attaining a length of 30–100 cm (10–40 in.) and a diameter of 15–50 cm (6–20 in.). Jackfruit can weigh up to 10–25 kg (22–55 lb) – the jack tree is another you shouldn't be caught napping under.

02 – Breadfruit

Artocarpus altilis

Derived from the wild species *Artocarpus camansi*, breadfruit is native to New Guinea, but domesticated varieties were taken to the South Pacific by early Austronesian migrants. This print is from *Bilderbuch für Kinder* (Picture Book for Children), 1790–1830, published by Friedrich Justin Bertuch.

Similar in appearance to the jackfruit is a further well-known but minor fruit: the durian (*Durio zibethinus*), native to Borneo and Sumatra. Durian fruit, like the jackfruit, is large and spiky, growing up to 30 cm (1 ft) long and 15 cm (6 in.) in diameter. Unlike the jackfruit, which comes from the fig family (*Moraceae*), the durian belongs to the mallow family (*Malvaceae*). The durian's notoriety comes from its punchy taste and smell. The British naturalist Alfred Russel Wallace famously described the taste of the durian:

> Durian fruit, like the jackfruit, is large and spiky, growing up to 30 cm (1 ft) long and 15 cm (6 in.) in diameter.

'The five cells are silky-white within, and are filled with a mass of firm, cream-coloured pulp, containing about three seeds each. This pulp is the edible part, and its consistence and flavour are indescribable. A rich custard highly flavoured with almonds gives the best general idea of it, but there are occasional wafts of flavour that call to mind cream-cheese, onion-sauce, sherry-wine, and other incongruous dishes. Then there is a rich glutinous smoothness in the pulp which nothing else possesses, but which adds to its delicacy. It is neither acidic nor sweet nor juicy; yet it wants neither of these qualities, for it is in itself perfect. It produces no nausea or other bad effect, and the more you eat of it the less you feel inclined to stop. In fact, to eat Durians is a new sensation worth a voyage to the East to experience.'

02

Although the taste of durian is agreeable to most people, the same cannot be said for the smell. Described variously as reminiscent of raw sewage, stale vomit, onions and turpentine, the stench has led to a ban on carrying durian fruits on public transport in places such as Singapore. Some years ago, we held a public event at Kew Gardens where visitors were given the opportunity to taste various fruits and vegetables that they may not have encountered before. For the durian tasting, we set up the stall in a separate tent some distance from the Orangery where the other fruits were on offer, and invited visitors to write down what they thought the durian tasted like. None of the descriptions were alike due to the complex chemicals involved and the associations they elicited on the tongue, though most people enjoyed the taste but were repelled by the smell.

03

What Makes Fruit Attractive to Eat?

Given their main purpose is to enable seed dispersal, fruits have evolved to attract animals – creatures that will eat their flesh and then excrete their seeds intact. While the flesh of the fruit is disposable to a tree, the seeds are vital for reproduction and must be protected. To this end, the seeds themselves need to be inaccessible or repellent – for example, to a seed-eating bird like a parrot. This is the case with the durian or breadfruit, where the seeds are encased by a large amount of sticky, glutinous flesh. Fruits also need to be off-putting if they are unripe, as the seeds they contain will be immature and not ready for dispersal. Examples of unripe fruits that are poisonous include the lychee (*Litchi chinensis*), which contains the toxin Hypoglycin A and causes severe 'hypoglycaemia' (low blood sugar) if large quantities of young fruit are eaten. Another example is the toxin urushiol, which is found in unripe cashew and mango fruits and causes severe allergic reactions in most people.

> Birds' eyesight is better than their sense of smell and, for these reasons, it is believed that birds are attracted by the colour of fruits.

03 – Wagtail with Fruit
Jan Brandes, 1785
Watercolour.

04 – Apple
Malus domestica
Bright colours indicate that
the fruit is ripe and ready to
eat, as well as being visible
from long distances.

04

The extent to which fruit colour attracts dispersers is contentious, but there is some evidence to suggest that colour is important to birds. Birds are 'tetrachromatic', meaning they have four different types of cone cells in the retinas of their eyes (compared to our two) and can therefore see a much wider range of colours than we can – including those on the ultra-violet spectrum. Birds' eyesight is better than their sense of smell and, for these reasons, it is believed that birds are attracted by the colour of fruits (particularly reds, purples and black), while mammals are more likely to be attracted by smell. Birds seem to actively avoid green fruits and there is some suggestion that they also don't like white fruits, although this appears to be anecdotal rather than based on scientific evidence.

Despite our comparatively limited perception of colour, we humans also prefer brightly coloured fruit because we associate their vibrancy with ripeness. Colour is therefore considered an important trait to confer when breeding fruit cultivars. Other valuable characteristics include disease resistance, uniformity of shape and texture, good storage potential and a pleasant odour and taste. Unfortunately, due to mass production and long global supply chains, modern fruit cultivars are often selected for their disease resistance, shelf life and uniformity rather than for their taste. As an example, apples have been cultivated for thousands of years, and the domesticated apple (*Malus domestica*) is primarily derived from four wild species (*M. sieversii, M. orientalis, M. sylvestris,* and *M. prunifolia*). Through the process of domestication, thousands of apple cultivars [pg.224] have been bred by humans, and today, we theoretically have access to more than 7,500 different kinds of apple. However, only about thirty cultivars (0.1 per cent of apple diversity) are traded commercially. The most popular apple varieties in the world are the Red Delicious, Gala, Granny Smith, Golden Delicious, Lady, Baldwin, McIntosh, Honey Crisp, Fuji and Cortland. These all have good shape uniformity, storage properties, texture and colour, and are easy to harvest, but there are certainly tastier varieties available.

One of the major challenges with breeding fruit cultivars is that many fruits are not 'true to type', meaning that if they are grown from seed, they won't necessarily have the same characteristics as their parent tree. For example, apples, pears, most peaches, some plums, apricots and cherries all need to be vegetatively propagated to ensure that the genetics of the parents will be preserved. Vegetative propagation usually entails taking cuttings or buds from the desirable fruit-producing parent tree (the scion) and grafting it on to the 'rootstock' of a different cultivar selected to grow vigorously in a certain soil type or climate. If you buy a fruit tree from a nursery you will see a characteristic bulge 7–8 cm (3 in.) above the root ball where the graft has taken.

Storing tree seeds at low temperatures in a seed bank [pg.22] is much cheaper than maintaining living trees but because fruit-tree seeds are not true to type, parent-tree tissue needs to be stored instead. Consequently, alternative technologies – such as the 'cryo-preservation' of buds or reproductive tissues in liquid nitrogen at ultra-low temperatures – are used to maintain fruit cultivars. Fruit collections of global importance include the UK's National Fruit Collection of apple cultivars maintained at Brogdale in Kent by the University of Reading and the global collection of breadfruit cultivars held by the Breadfruit Institute at the National Tropical Botanical Garden in Hawai'i. Perhaps the most surprising location of a global fruit collection is the International Musa Germplasm Transit Centre (ITC) in Leuven, Belgium, which is home to the world's largest collection of banana cultivars.

> If you buy a fruit tree from a nursery you will see a characteristic bulge 7-8 cm (3 in.) above the root ball where the graft has taken.

05 – The International Musa Germplasm Transit Centre
The cold storage room for the banana collection at this facility in Belgium. Because many fruit trees do not pass on the desirable characteristics of their fruit via seed, they are cloned from cuttings in laboratories in a process called micro-propagation.

Commercial fruit orchard
Cloned fruit cultivars ensure consistency in fruit traits such as colour, taste, texture and storability – all important to customers and retailers.

05

Mutiny on the Bounty

In April 1789, the British HMS *Bounty* was seized by disgruntled crewmen during its mission to transport breadfruit from Tahiti to the West Indies. Led by Fletcher Christian, the mutineers set their captain, William Bligh, along with eighteen loyalists, adrift in an open launch. Christian and several of his followers went into hiding on Pitcairn Island, where descendants of the surviving mutineers still live today.

← The Mutineers Turning Lieutenant Bligh and Part of the Officers and Crew Adrift, 29th April 1789
Robert Dodd, pub. 1790
Hand-coloured engraving.

↓ Breadfruit (*Artocarpus altilis*)
Ebenezer Sibly, c. 1798
From *A Key to Physic*.
The breadfruit transported on the HMS *Bounty* takes its name from its white pulp, thought to resemble new bread.

Dodd del. *The Bread Fruit Tree.* *Pratten sculp.*

Size and Weight

Tree fruits come in all kinds of shapes and sizes. The largest and heaviest tree fruit in the world is the jackfruit (*Artocarpus heterophyllus*), with the African sausage tree (*Kigelia africana*) not far behind.

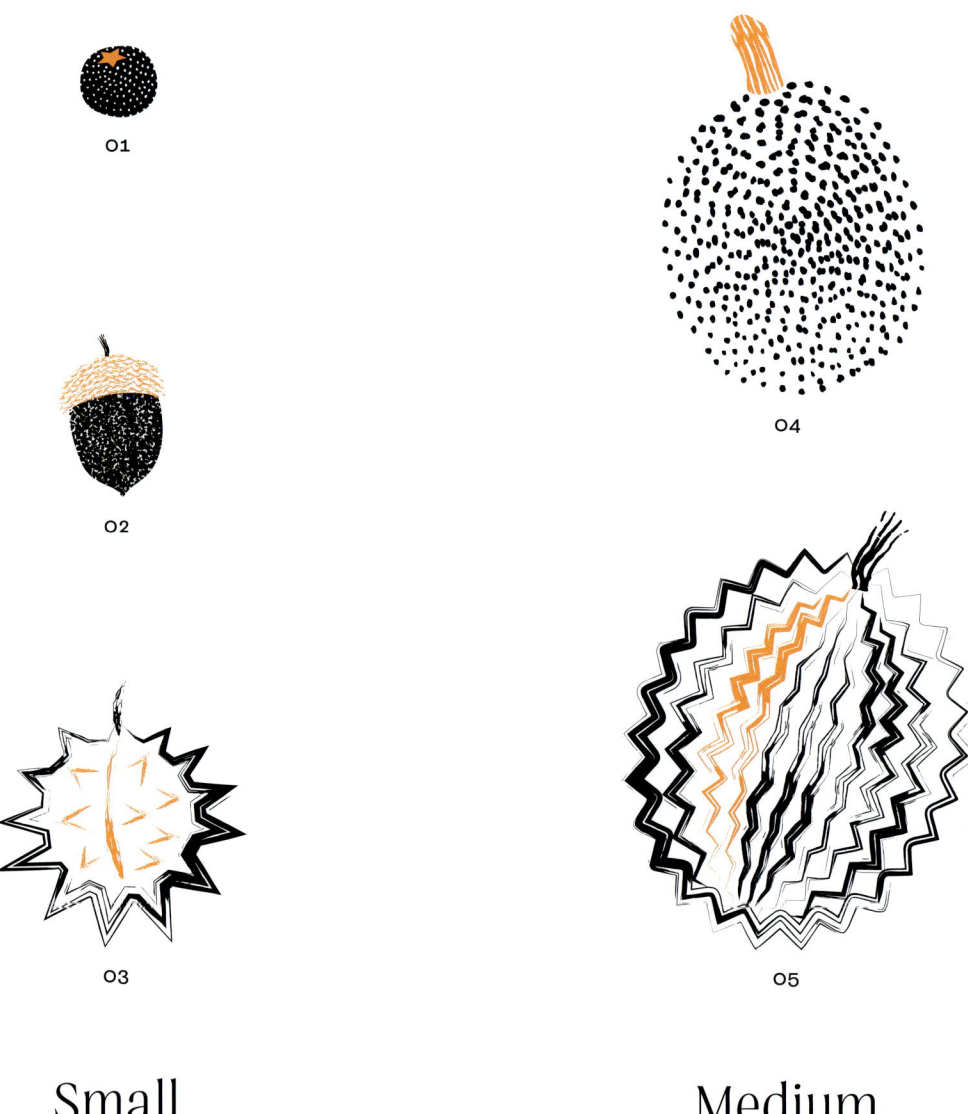

01

02

03

04

05

Small

Not to scale

Medium

Not to scale

01 – Hawthorn
Crataegus monogyna
Size: 4 mm (0.15 in.)
Weight: 0.1 g (0.003 oz)

02 – English oak
Quercus robur
Size: 2.7 cm (1 in.)
Weight: 2.5 g (0.088 oz)

03 – Horse chestnut
Aesculus hippocastanum
Size: 5 cm (2 in.)
Weight: 13 g (0.5 oz)

04 – Breadfruit
Artocarpus altilis
Size: 20 cm (7.8 in.)
Weight: 4 kg (8.8 lb)

05 – Durian
Durio zibethinus
Size: 30 cm (1 ft)
Weight: 4 kg (8.8 lb)

06 – Sausage tree
Kigelia africana
Size: 60 cm (2 ft)
Weight: 22 kg (3.5 stone)

07 – Jackfruit
Artocarpus heterophyllus
Size: 90 cm (3 ft)
Weight: 55 kg (8.5 stone)

06

07

Large

Not to scale

Most Popular Fruits by Production

Tree fruits are amongst the most popular fruits traded commercially. Bananas take the number one spot, while apples and oranges are also in the top five.

East Indian Market Stall in Batavia
Albert Eckhout, 1640–66
Oil on canvas.

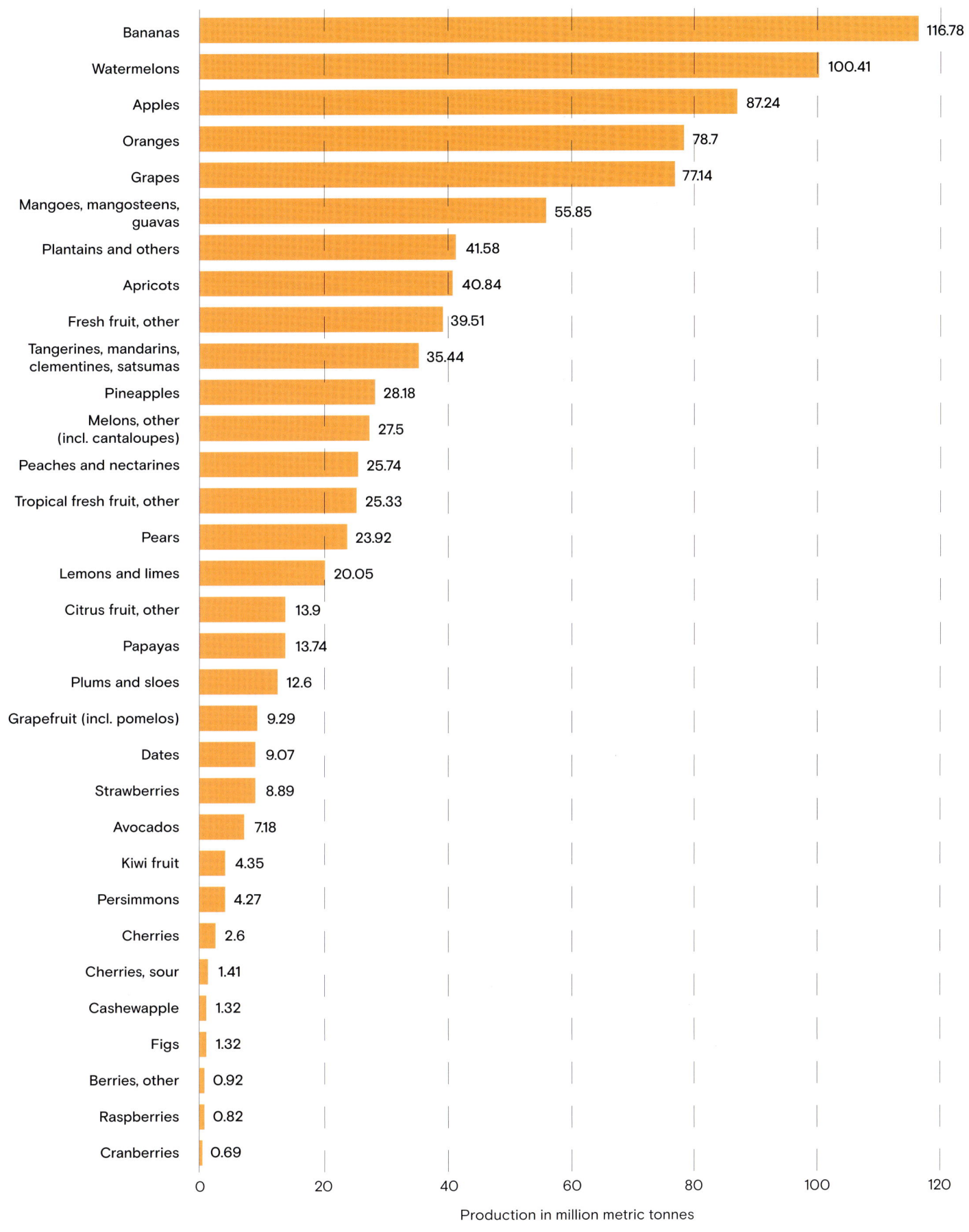

Fruit	Production in million metric tonnes
Bananas	116.78
Watermelons	100.41
Apples	87.24
Oranges	78.7
Grapes	77.14
Mangoes, mangosteens, guavas	55.85
Plantains and others	41.58
Apricots	40.84
Fresh fruit, other	39.51
Tangerines, mandarins, clementines, satsumas	35.44
Pineapples	28.18
Melons, other (incl. cantaloupes)	27.5
Peaches and nectarines	25.74
Tropical fresh fruit, other	25.33
Pears	23.92
Lemons and limes	20.05
Citrus fruit, other	13.9
Papayas	13.74
Plums and sloes	12.6
Grapefruit (incl. pomelos)	9.29
Dates	9.07
Strawberries	8.89
Avocados	7.18
Kiwi fruit	4.35
Persimmons	4.27
Cherries	2.6
Cherries, sour	1.41
Cashewapple	1.32
Figs	1.32
Berries, other	0.92
Raspberries	0.82
Cranberries	0.69

Production in million metric tonnes

Apple Varieties

While over 7,500 varieties, or cultivars, of the domestic eating apple (*Malus domestica*) are known, the vast majority are not suitable for mass production.

McIntosh
The McIntosh has a red and green skin, a tart flavour, and tender white flesh.

Braeburn
Braeburn apples have a combination of sweet and tart flavour and originate from New Zealand.

Honeycrisp
This apple's sweetness, firmness, and tartness make it an ideal apple for eating raw.

Empire
Introduced in 1966, the Empire has white subacid flesh, a tangy taste and a ruby-red colour.

Red Delicious
The Red Delicious originated at an orchard in Iowa in 1872 as 'a fruit of surpassing sweetness'.

Fuji
Developed in Japan in the 1930s, Fuji is a dark-red, conic apple. Its sweet, crisp, dense flesh is very mildly flavoured.

Gala

Thin, tannic skin is yellow-green, with a red blush overlaid with reddish-orange streaks. Flesh is yellowish-white, crisp and grainy, with a mild flavour.

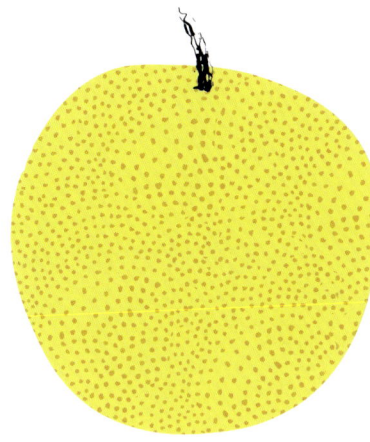

Golden Delicious

One of the most popular varieties in the world. Uniform light green-yellow coloration. Flesh firm, crisp, tender, juicy, mild subacid, aromatic.

Granny Smith

A favourite variety, widely sold in the UK. Lime-green colouring. Extremely tart. Popular for pies.

Pink Lady (Cripps pink)

This sweet-tart apple has both high sugars and acids, with a crisp bite and effervescent finish.

Opal

Firm, fine to medium grained, medium juicy, full flavoured, sweet, mild subacid.

Piñata

Piñata apples were created by researchers in Dresden-Pillnitz, Germany in the 1970s as a cross between three heirloom apples: Golden Delicious, Cox's Orange Pippin, and Duchess of Oldenburg.

Health Benefits

The health benefits of tree fruits are well known. They are an excellent source of vitamins, micronutrients and fibre. They also contain a wide range of antioxidants, including flavenoids, which boost the immune system.

1 lemon (*Citrus limon*) contains:

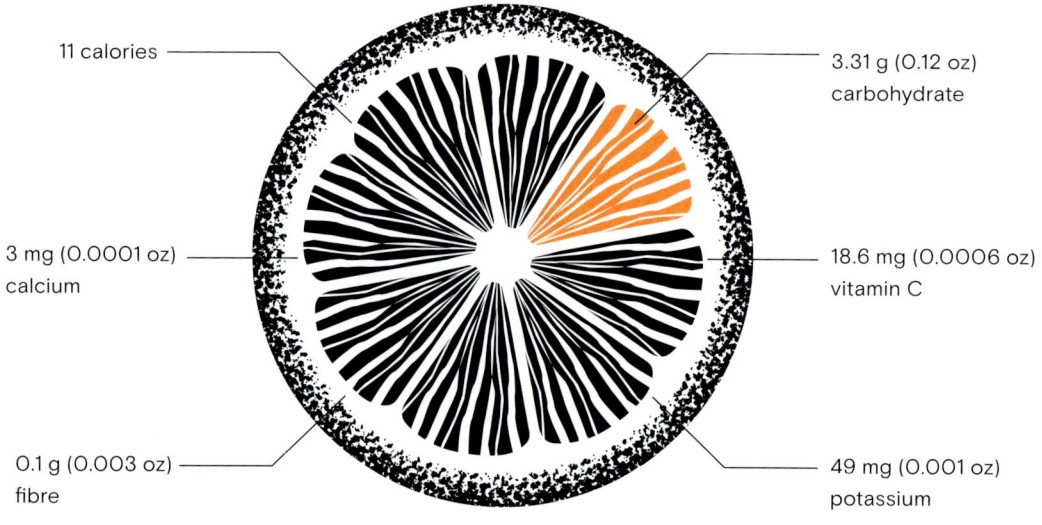

11 calories

3.31 g (0.12 oz) carbohydrate

3 mg (0.0001 oz) calcium

18.6 mg (0.0006 oz) vitamin C

0.1 g (0.003 oz) fibre

49 mg (0.001 oz) potassium

1 apple (*Malus domestica*) contains:

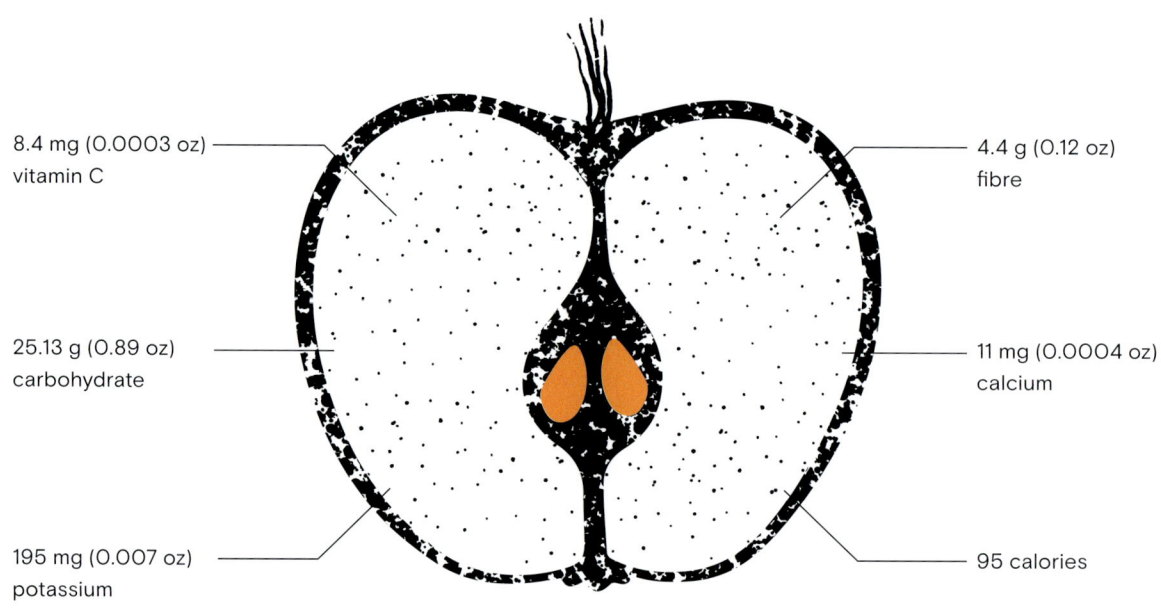

8.4 mg (0.0003 oz) vitamin C

4.4 g (0.12 oz) fibre

25.13 g (0.89 oz) carbohydrate

11 mg (0.0004 oz) calcium

195 mg (0.007 oz) potassium

95 calories

1 pomegranate (*Punica granatum*) contains:

234 calories

52.73 g (1.86 oz) carbohydrate

11.3 g (0.4 oz) fibre

666 mg (0.02 oz) potassium

28 mg (0.001 oz) calcium

28 mg (0.001 oz) vitamin C

1 banana (*Musa*) contains:

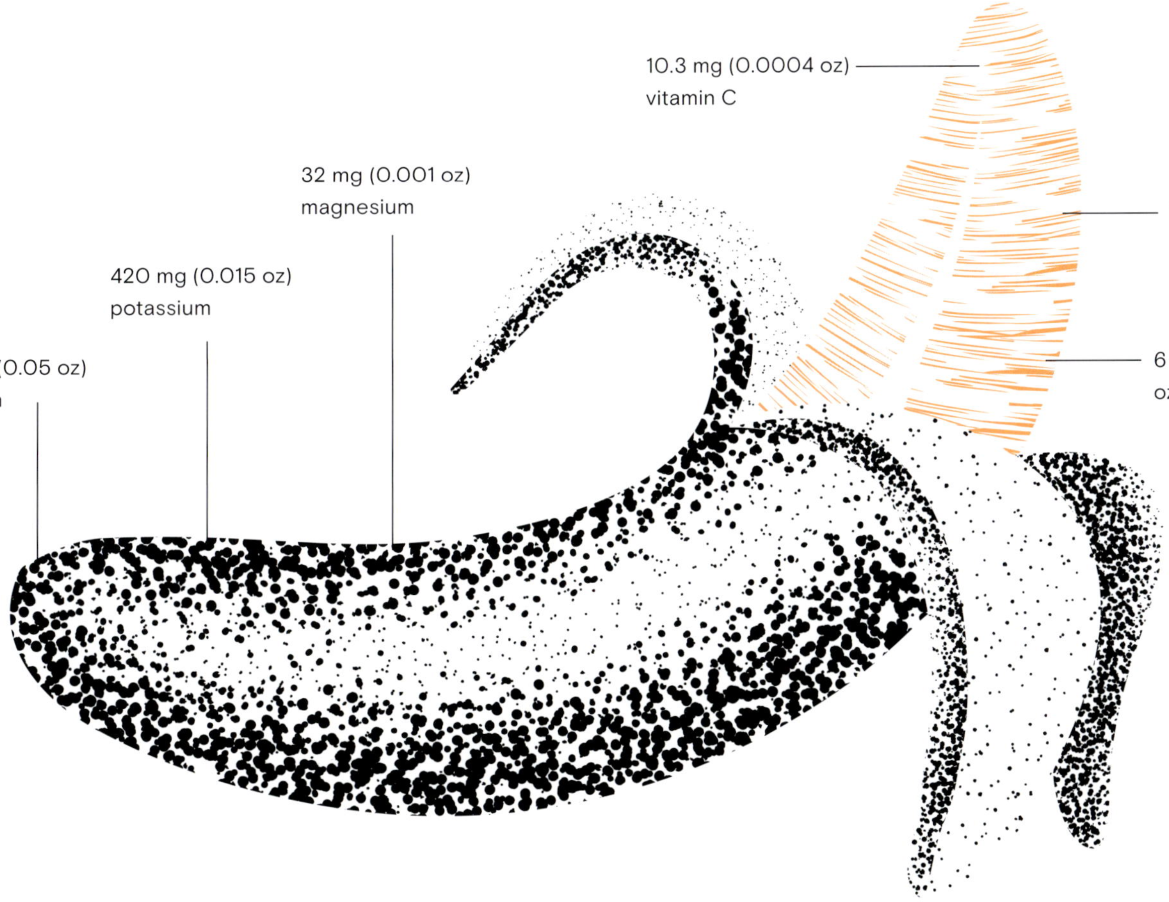

10.3 mg (0.0004 oz) vitamin C

32 mg (0.001 oz) magnesium

130 calories

420 mg (0.015 oz) potassium

1.29 g (0.05 oz) protein

6 mg (0.0002 oz) calcium

Fruits ⟶ Health Benefits ⟶ Citrus limon, Malus domestica, Punica granatum, Musa

Fruit-based Fashion

Orange Fiber is the world's first brand to produce sustainable fabrics from citrus juice by-products. The innovative process was patented in 2014 in citrus-producing countries all over the world.

← Orange Fiber fabric
Salvatore Ferragamo was the first fashion house to use Orange Fiber fabrics. Their capsule collection was created to celebrate Earth Day 2017, with sustainable Orange Fiber materials adorned with prints by Mario Trimarchi.

↓ Salvatore Ferragamo's capsule collection successfully married cutting-edge processes with timeless Italian design. To the left, the honeybell orange (*Minneola tangelo*), is an innovation in its own right. It isn't technically an orange at all, but a cross between a tangerine and a grapefruit.

Fruit Juices and Beverages

Tree fruits are not only eaten – they are also drunk. Between 2019 and 2020, 1.62 million metric tonnes (1.6 imperial tons) of orange juice were produced – equivalent to more than 1 billion litres (0.2 billion gallons) – the majority coming from Brazil, the world's largest producer of oranges. Apple juice is not far behind, the leading producer being China. Of course, apples are not solely used to make fresh apple juice; they are also the basis of the alcoholic beverage, cider. France is the world's top producer of cider, while the United Kingdom is the greatest consumer per capita. There is some evidence that the Celts made cider from crab apples as early as 3000 BCE, while the arrival of the Romans in Britain brought developments in orchard techniques and new apple cultivars to British shores. The Norman Conquest, effected by the Battle of Hastings in 1066, later introduced more acidic apple varieties to Britain, which were better suited to making cider. Normandy and Brittany remain the centre of cider-apple diversity to this day. Whereas European cider (or *cidre* in French) typically comprises 5–12 per cent alcohol, in the USA and parts of Canada, the term 'cider' (as opposed to 'hard cider') applies to non-alcoholic apple juice.

The apple-based beverage Calvados, meanwhile, is double-distilled cider with an alcohol content of around 40 per cent. Other 'hard liquors' derived from tree fruits include schnapps, palinkas and brandies. Here, the drink is produced from the fruit mash, or through mixing fruit with grain spirit, to create a strong liqueur. The most common fruit brandies are made from peaches, apricots, pears, plums and cherries.

> Between 2019 and 2020, 1.62 million metric tonnes (1.6 imperial tons) of orange juice were produced – equivalent to more than 1 billion litres (0.2 billion gallons).

In one of my favourite short stories by early 20th-century South African writer Herman Charles Bosman, who gently satirized Dutch Afrikaaner farmers in the Groot Marico District, the main protagonist, an old farmer named Oom (Uncle) Schalk Lourens, shares a mule-and-cart journey with the local church minister one very cold winter night. Desperate to gain access to the bottle of peach brandy he keeps in the back of the cart, but knowing the preacher would disapprove, he explains to the minister that the mules are getting tired. Luckily, he has just the remedy – a medicinal spirit that he delivers by blowing into the beasts' nostrils. Immediately perking up, the preacher volunteers to carry out this onerous task himself, explaining that it's his job to administer to the needs of *all* God's creatures.

Still on the continent of Africa, and ending this chapter as we started, one of the most recent liqueurs developed from a relatively obscure fruit is 'Amarula'. This distinctively flavoured beverage is distilled from the fruits of the southern African marula tree (*Sclerocarya birrea*), which is related to the mango but has much smaller fruits. The fruit is about the size of a plum, with a thin, sweet flesh around a single stone. Marulas are a great favourite of elephants – which consume them in vast quantities – and other wildlife. If left on the ground for a week or two, the fruits ferment without any human intervention. In Jamie Uys's 1974 film *Animals are Beautiful People*, elephants, warthogs and baboons were all documented very much the worse for wear after over-indulging in marula fruits.

The Mazagon Mango of Bombay, with the Papilio Bolina Purple-eyed Butterfly
W. Hooker after J. Forbes, 1768 (pub. c. 1813)
Coloured aquatint. Mango (*Mangifera indica*) flower and fruit with a purple-eyed butterfly (*Papilio*).

The MAZAGON MANGO *of Bombay,*
with the PAPILIO BOLINA *Purple-eyed Butterfly*

Jam. Forbes, 1768

Fruit and Alcohol

Alcoholic beverages derived from fruit have probably been around for as long as humankind, and it is easy to see how they may have come about. Fermentation is a natural process through which yeasts convert sugars to alcohol.

01

02

01 – Cherry Distillery, Eiken, Switzerland
Black and white photograph, 1949
Fruits must first be pressed to separate the juice from the pulp and seeds.

02 – Process
At the same Swiss distillery, cherries are being cold-pressed, after which the juice will be pasteurized, strained and filtered.

03 – Oranje Bitter
This vintage label was created for a Dutch bitter orange liqueur. 'Het Leeuwke' means 'The Lion'.

04 – Cherry brandy
Cherry brandy can also be made by steeping cherries in a neutral spirit like vodka.

05 – Cherry liqueur
Cherry liqueur is more syrupy and sweet than cherry brandy, as it has added sugar. Brandy is more common so, confusingly, cherry liqueur is often referred to as cherry brandy.

03

04

05

Architecture

Architects may be genuinely inspired by the shapes of fruit or, perhaps more frequently, the public might interpret the allusion for them. The Esplanade Theatres on the Bay in Singapore, consisting of two buildings, a concert hall and a theatre, is colloquially known to Singapore locals as 'The Big Durians', although this wasn't originally the architects' intention. Earlier, less-flattering comparisons were made to 'two copulating aardvarks'.

← Esplanade Theatres on the Bay, Singapore
DP Architects and Michael Wilford & Partners, 2002
More than 7,000 triangular aluminium sunshades that cover its two circular glass shell structures look somewhat like spikes on two halves of a durian fruit.

↓ Durian fruit
Durio zibethinus
The durian is banned from public transport in Singapore because of its unpleasant smell.

Art

Fruit has been depicted in illustrations and works of art for millennia, often imbued with symbolism. Ripe fruit frequently represents life, vitality and fertility, while its eventual decay signifies human transience and mortality. The fruit and other foodstuffs painted in Egyptian tombs, meanwhile, were believed to become tangible in the afterlife.

01

01 – Fruit Trees and Herbs
in Java
Anon., 1646
Etching.

02 – Mango
*Michaeł Boym, Flora
Sinensis, 1656*
Book illustration.

03 – Mangosteen
Fruits series (N12), 1891
Trade card from a set of
fifty, issued to promote
Allen & Ginter cigarettes.

02

03

Symbiosis

○ Symbiosis

'Symbiosis' is one of those scientific terms that has entered into everyday use. It refers to the interaction between two distinct organisms – a relationship that can be beneficial to one actor or both. Over time, trees have developed symbiotic relationships with a huge range of different plants, fungi and animals, including humans. Understanding the intricate web of species that depend on trees shows just how vital they are to our ecosystems and our planet.

01

The word symbiosis comes from the Greek 'symbiōsis', meaning 'living together'. This is generally what we understand symbiosis to mean today, though we might put slightly more emphasis on co-dependence. Scientists like to complicate matters, however, and have defined various kinds of symbiosis according to those that benefit and those who don't. 'Mutualism' describes a situation in which all parties benefit from the relationship; 'commensalism' refers to an interaction in which only one party benefits, but in a non-detrimental way to the other; and 'parasitism' involves one organism benefiting at the expense of another. Rather like modern families, trees engage in all kinds of complicated relationships, including these three forms of symbiosis.

The relationship between many mycorrhizal fungi and their tree hosts is mutualistic, as discussed in Chapter 3 [pg.88]. Mycorrhizal mutualism involves a fungus colonizing a tree's root system. In this exchange, tree and fungi roots are connected beneath the soil: trees benefit from access to water and hard-to-get nutrients like phosphates absorbed by the fungus, while the fungus gains sugars and other nutrients from the tree. The most common kind of mycorrhizal symbionts of trees are 'ectomycorrhizal' fungi, so called because the thread-like filaments of the fungi form an external sheath around the roots of the tree, exchanging water and nutrients through the root surface. Most familiar mushrooms and toadstools are ectomycorrhizal symbionts of trees.

The famous red and white 'toadstool' fly agaric (*Amanita muscaria*), for example, forms symbiotic relationships with many trees, including oak, fir, pine, spruce, cedar and birch. Fly agaric is infamous for its hallucinogenic properties, but many of our edible mushrooms, including boletes and chanterelles, are also ectomycorrhizal symbionts. In south-central Africa, edible mushrooms found in miombo woodlands form an important part of local diets, some of them growing to an impressive size. The largest edible mushroom in the world, the wonderfully named *Termitomyces titanicus*, can grow up to a metre (3 feet) in diameter. In this case, the symbiosis is actually between the mushroom and termites, which create the ideal temperature for mushroom growth within their elaborately designed nests; in return, the fungus breaks down tree wood that would otherwise be indigestible to the termites.

Another kind of tree mutualism involving microorganisms is the relationship that certain kinds of tree in the legume (pea) family have with nitrogen-fixing 'rhizobia' bacteria. These are called 'endosymbionts' because the tree hosts the bacteria internally in root nodules. In exchange for its hospitality, the tree is provided with nitrogen compounds essential for tree growth.

Commensalism is much more common than mutualism because so many different species live off trees without causing them any harm. Trees provide a substrate (a base on which an organism grows) for a huge range of other plants, particularly in the tropics. Plants that live on other plants are called 'epiphytes'. Some are parasitic (see below), but epiphytes such bromeliads, orchids and ferns [pg.256] simply use the tree as a solid surface on which to perch. 'Air plants' like Spanish moss (*Tillandsia usneoides*) have no need for aerial roots or soil, instead obtaining all the moisture and minerals they need from rainfall and the air.

> The famous red and white 'toadstool' fly agaric (*Amanita muscaria*) [...] forms symbiotic relationships with many trees, including oak, fir, pine, spruce, cedar and birch.

Lichens, mosses [pg.258] and liverworts may also be epiphytic, and lichens, in particular, are associated with almost all trees. One of these, *Usnea*, looks very like Spanish moss and is sometimes called 'old man's beard' or 'beard lichen' because of the way it drapes across tree branches. In some trees, such as California's redwoods (*Sequoia sempervirens*), there are hundreds of different epiphytes growing; these include vascular, rooting trees and shrubs that grow in the organic matter that accumulates in the nooks and crannies of redwoods up to 100 metres (300 feet) above the forest floor. Single redwood trees can form an ecosystem all of their own.

Mistletoes, although parasitic, usually don't actually do the tree a great deal of harm. It's only when a tree is weakened by other causes that an infestation can occur, leading to death.

02

Some of the best-known epiphytes are parasitic. The mistletoe (*Viscum album*) that we hang up at Christmas is a hemi-parasite, meaning that it can photosynthesize and make some of its own food, but that it mainly draws water and nutrients from the host tree via a root-like projection called a 'haustorium' that grows into the host. The largest family of epiphytic parasites are the *Loranthaceae*, and these are mainly found in the southern hemisphere. Unlike *Viscum*, the *Loranthaceae* produce beautiful, long tubular flowers with bright colours. Their fruits are eaten and dispersed by birds, who excrete the sticky seeds onto tree branches, from where they germinate. Many years ago, when I was a young botanist working in the Muchinga Mountains of Zambia, a *Loranthus* was pointed out to me in the crown of a host tree. I was told that its name was 'Mpumbamakoa' – the 'thing that is out of place'. My guide, an old Zambian herbalist, grinned and explained that the same expression was sometimes used to describe Englishmen wandering about in the forest.

03

O4

Mistletoes, although parasitic, usually don't do the tree a great deal of harm. It's only when a tree is weakened by other causes that an infestation can occur, leading to death. Other parasitic plants are more fatal to their hosts. Most notable amongst these are the strangler figs. Also deposited by birds, strangler-fig seeds germinate in the organic detritus of the crook of a tree, from where they send down aerial roots to the forest floor. From then on, they slowly but surely outcompete the host tree, strangling and depriving it of light, water and nutrients from the soil. An example of a strangler fig is the banyan tree (*Ficus benghalensis*), India's national tree. The name 'banyan' comes from the 'Baniya' traders who rest under the shade of these great trees.

Risky Relationships

As described above, mutualistic relationships benefit both parties, but the consequences can be severe if one of those partners gets into trouble. Screw pines (*Pandanus*, pg.252), for example, are an important component of Madagascar's humid and semi-humid forest, but have come increasingly under threat in a country where over 4 million hectares (10 million acres) of primary forest has been lost due to clearing and fire damage in the past twenty years. Screw pines provide a habitat for a wide range of animals in the water they collect in cavities formed by their leaves ('phytotelmata'). In one well-known study, 20 species (9 frogs, 6 geckoes, 4 snakes and 1 skink) were found in these plants; at least 5 species were identified as obligate *Pandanus*-dwellers, meaning that they are only found in *Pandanus* plants. Other studies suggest that at least 21 additional species in Madagascar are commonly or exclusively found in *Pandanus* plants. This means that if the *Pandanus* tree becomes extinct, then so do many other species.

These 'obligate' relationships are not confined to the tropics. The recent drastic decline in the European ash (*Fraxinus excelsior*) due to the introduction of a fungal pathogen that causes ash dieback is having a major impact on its symbionts as well. Populations of host-specific organisms – such as the tabby knot horn moth (*Euzophera pinguis*), the scarce fritillary butterfly (*Euphydryas maturna*), the jewel beetle (*Agrilus convexicollis*), the epiphytic moss *Neckera pennata* and fungi such as *Pyrenula nitidella* and *Perenniporia fraxinea* – are all likely to fall significantly. Of course, these organisms will be integral to the life cycles of numerous other plants and animals too, creating widespread repercussions that scientists call the 'ecological cascade effect'. Because trees provide a habitat and food source for so many other organisms, they are referred to as 'keystone species' – if they are lost, the whole building could come crashing down.

Why Tree Extinctions Matter

Human dependence on trees is explored more fully in Chapter 9 [pg.270], but our knowledge of how trees and plants underpin the life systems of the planet is still incomplete. The 2021 'State of the World's Trees' report warns that more than 17,500 species of tree (about 30 per cent of all tree species) are threatened with extinction [pg.264]. The report also indicates that 142 tree species are known to be extinct, with a further 440 species on the brink of extinction with fewer than 50 individuals remaining in the wild.

> Species that are very close to extinction include *Hyophorbe amaricaulis* in Mauritius – known as the loneliest tree in the world.

Species that are very close to extinction include *Hyophorbe amaricaulis* in Mauritius – known as the loneliest tree in the world. This single-remaining specimen in Curepipe Botanical Garden is 150 years old, and as far as we know is the only one of its kind. It is diseased and probably will not survive much longer. So far, all attempts to propagate it have failed. Another example from the International Union for Conservation of Nature (IUCN) Red List is *Robinsonia berteroi*, which is endemic to Robinson Crusoe Island – part of the Juan Fernández Island archipelago off the coast of Chile. There is currently one-known living individual, which sits on the top of El Yunque hill. According to records, this species was once abundant, and is thought to have declined significantly between 1908 and 1982 due to clearing for agriculture and development. *Robinsonia berteroi* was considered extinct after what was thought to be the last tree of its kind died in 2004, prior to the discovery of the El Yunque individual in 2015. Cultivation of this species is recommended for conservation.

05 – Potato fields
Clearing of natural vegetation for agriculture has changed the face of the landscape and is the single biggest threat to native plants and habitats.

Pleodendron costaricense is another species on the brink of extinction, a large tree endemic to Costa Rica, of which only two or three mature individuals remain. Although the trees are flowering and fruiting, little regeneration is evident, and two of the trees are rooted along roads built for dams and logging. Due to its small population size, the species is listed as Critically Endangered. *Hibiscus bennetti*, meanwhile, has a very small population in a single location on Vanua Levu in Fiji; it is likely that only two individuals remain post Cyclone Winston in 2016. This species – along with many others – is threatened by increasingly frequent and intense weather events, including cyclones, which can destroy both its habitat and existing mature individuals. A final example is *Croton leptanthus*, which is endemic to Lasanag Island, Morobe Province, Papua New Guinea. Only one tree of its kind remains on the island, where the species is at risk from the expansion of home gardens and human exploitation of the forest.

Clearly, the main drivers of extinction are our own activities – primarily land clearing for agriculture, forestry and human habitation – while the direct and indirect effects of climate change, such as the increasing frequency of storms, floods, fires and emerging pests and diseases, are having a growing impact. Human over-exploitation is a major cause of tree decline, and the 'State of the World's Trees' report indicates that one in five species threatened with extinction has a direct use to humans for food, medicine, construction and so on.

05

The world's botanic gardens and arboreta grow about 18,000 species of tree, but only around 500 species are widely cultivated for timber, medicines, fibre and fruit. For this reason, it's quite reasonable to ask why all these other tree species matter. Firstly, a huge number of species are harvested directly from the wild, with no cultivation (or replenishment) involved. For many of the world's poorest people, who rely on them for food, medicine and shelter, these trees emphatically matter. Secondly, the risk of the cascade effect mentioned above cannot be overstated. As keystone species, tree relationships with other organisms are so complex and far reaching that we simply don't know what knock-on effects will result from the loss of a single species. There may be no discernible effect on our lives or, like falling dominoes, the impacts could be catastrophic.

The understanding that we are part and symbionts of nature, not somehow above it, is well established in traditional cultures. And while it is difficult to pinpoint where the opposite idea – of man's dominion over nature – originated, it too goes back a long way. Genesis 1:28 says:

> 'God blessed them and said to them, "Be fruitful and increase in number; fill the earth and subdue it. Rule over the fish in the sea and the birds in the sky and over every living creature that moves on the ground."'

Much later, the notion of humans as 'masters and possessors of nature' (Descartes) gained currency during the Enlightenment, and accelerated with industrialization, advances in medical science and other new technologies. In 1992, the United Nations Convention on Biological Diversity actually

Lichens are epiphytes, meaning that they live on the surface of other plants. The lichen itself is a composite organism that arises from a mutualistic relationship between a fungus and a 'cyanobacteria' or alga.

enshrined in international law the sovereignty of nations over biodiversity. Up until this point, the generally accepted view was that nature existed for the common good, but the new framework established nature as effectively being 'owned' by countries. The rationale was that if states owned their biodiversity and were able to harness its financial value, they would also be more likely to look after it. In practice, neither of these things happened, and all kinds of legal problems have arisen since; nature pays little attention to political boundaries drawn on maps by people. Who owns transnational resources is one issue, but another is that we all benefit from each other's biodiversity. If a single nation decides to deny access to or demand reparations for its biodiversity – something that before 1992 was regarded as a common good – this could significantly disrupt food systems, access to medicines and so forth. To avoid this issue, various treaties supersede the Convention, including the rather unwieldy International Treaty on Plant Genetic Resources for Food and Agriculture (ITPGRFA), which enables nations to exchange plant material of the world's major crops without paying the country of origin.

Many people now wonder whether the Convention on Biological Diversity took a major wrong turn by codifying ownership of biodiversity. Ultimately, we are all symbionts of nature, though we may not always be aware of the various relationships at play or what's at stake. The extinction of any species can have profound effects – particularly if we are the cause.

Tree Symbionts

Tree symbiosis includes examples of mutualism (where all parties benefit), commensalism (where one party benefits but does no harm) and parasitism (where one party benefits but harms the other).

Fly agaric – Has a mutualistic relationship with its tree host, exchanging water, minerals and sugars with the tree.

Lichens – Have a commensal relationship with their hosts, using the tree as a substrate but doing no damage to it.

Bromeliads – Epiphytes that use the tree as a substrate in a commensal relationship that doesn't harm the host.

Mosses – Use trees as a substrate, usually growing on the shady or wetter side of the tree to catch maximum moisture.

Mistletoe – Hemi-parasites, photosynthesizing and making their own food but also deriving nutrients from the host tree.

Orchids – Epiphytic orchids grow in tree crooks or on their trunks or branches. They do no harm to the tree.

Screw Pine of Madagascar

Madagascar's humid and semi-humid forests are home to nearly ninety species of screw pine (*Pandanus*). These trees in turn provide a habitat for reptiles and amphibians in the water they collect in cavities formed by their leaves (phytotelmata).

→ Habitat provision
A well-publicized study found 20 reptile and amphibian species (9 frogs, 6 geckoes, 4 snakes and 1 skink) in the phytotelmata of a single Pandanus.

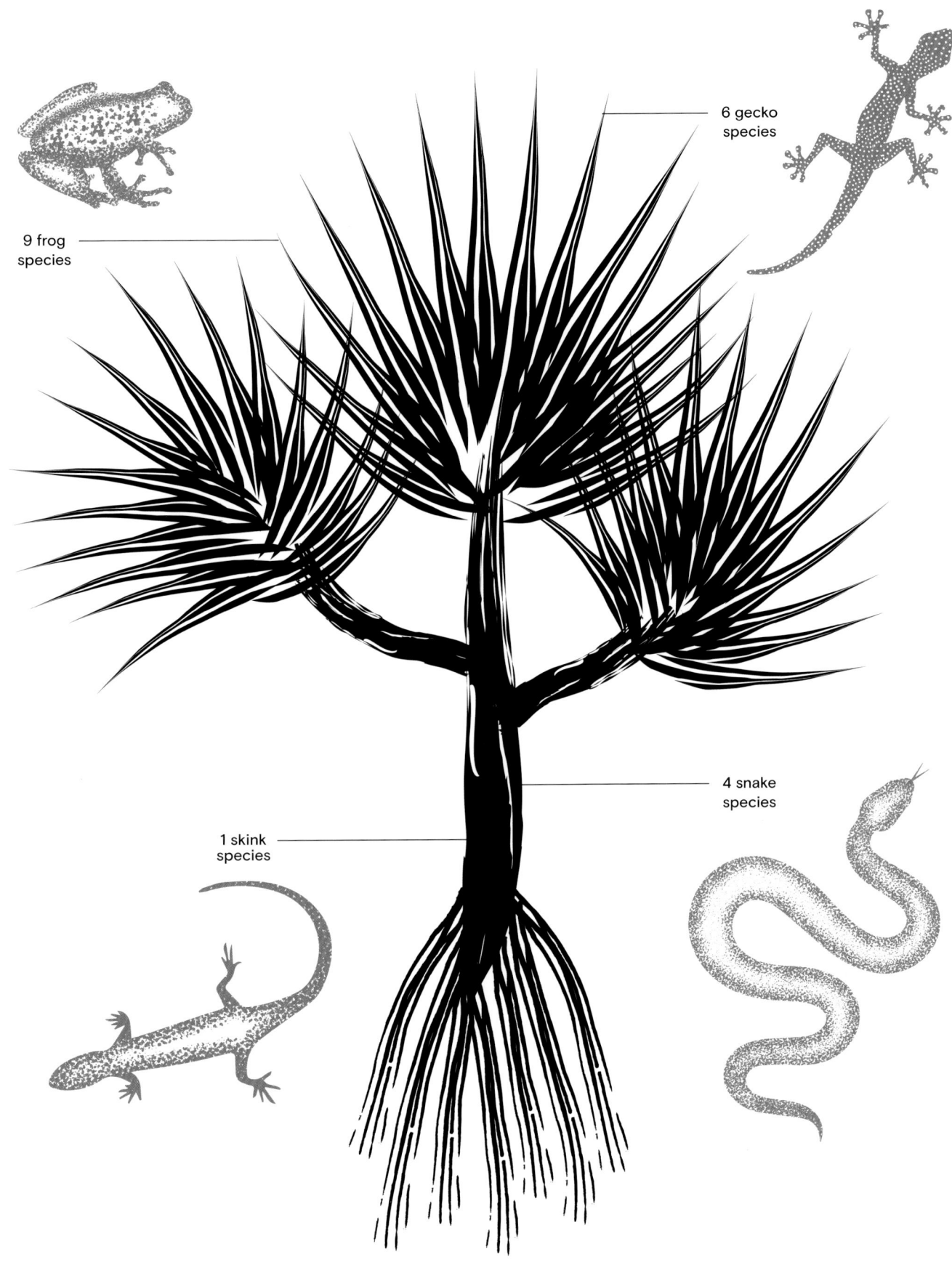

9 frog
species

6 gecko
species

1 skink
species

4 snake
species

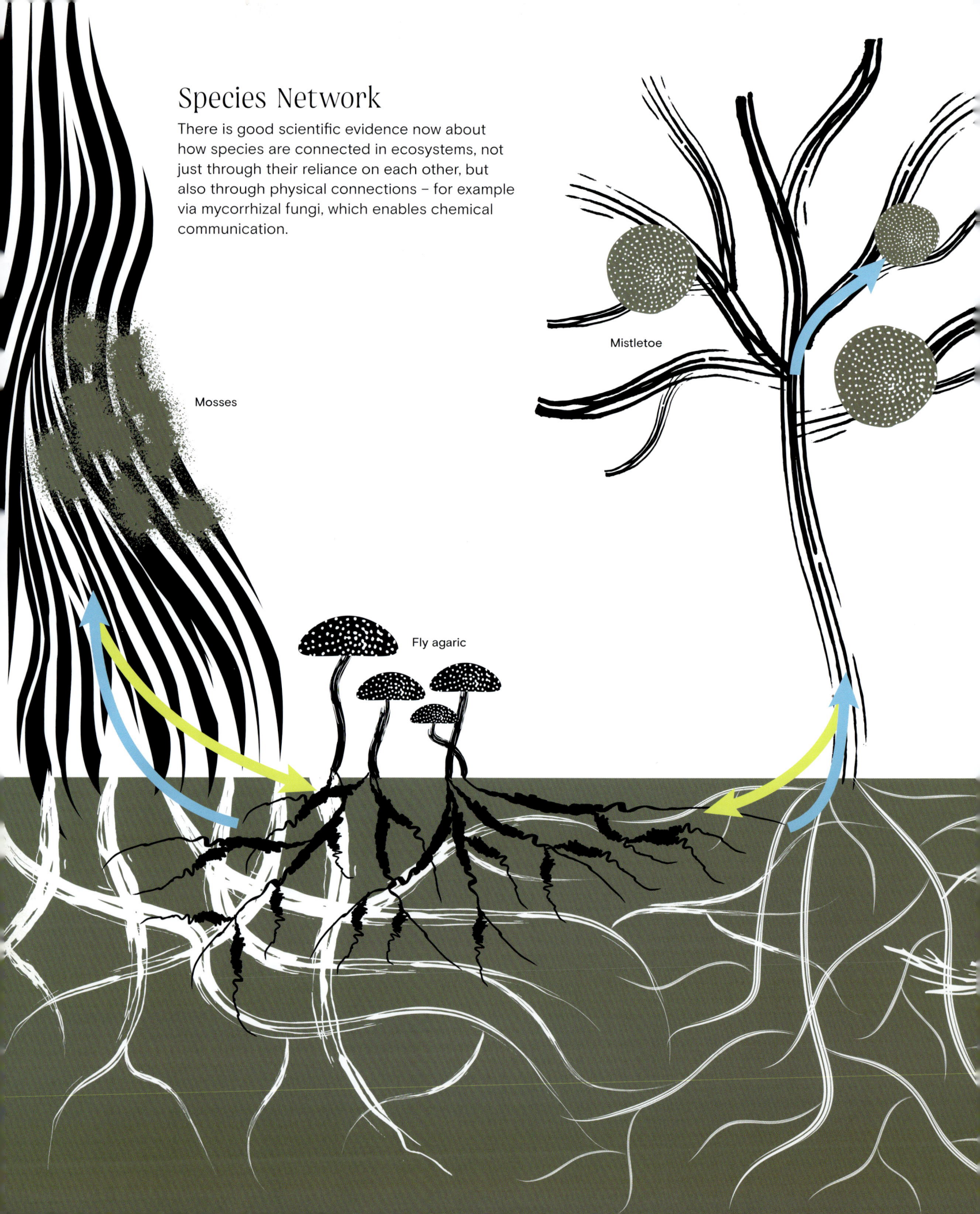

Species Network

There is good scientific evidence now about how species are connected in ecosystems, not just through their reliance on each other, but also through physical connections – for example via mycorrhizal fungi, which enables chemical communication.

Mosses

Mistletoe

Fly agaric

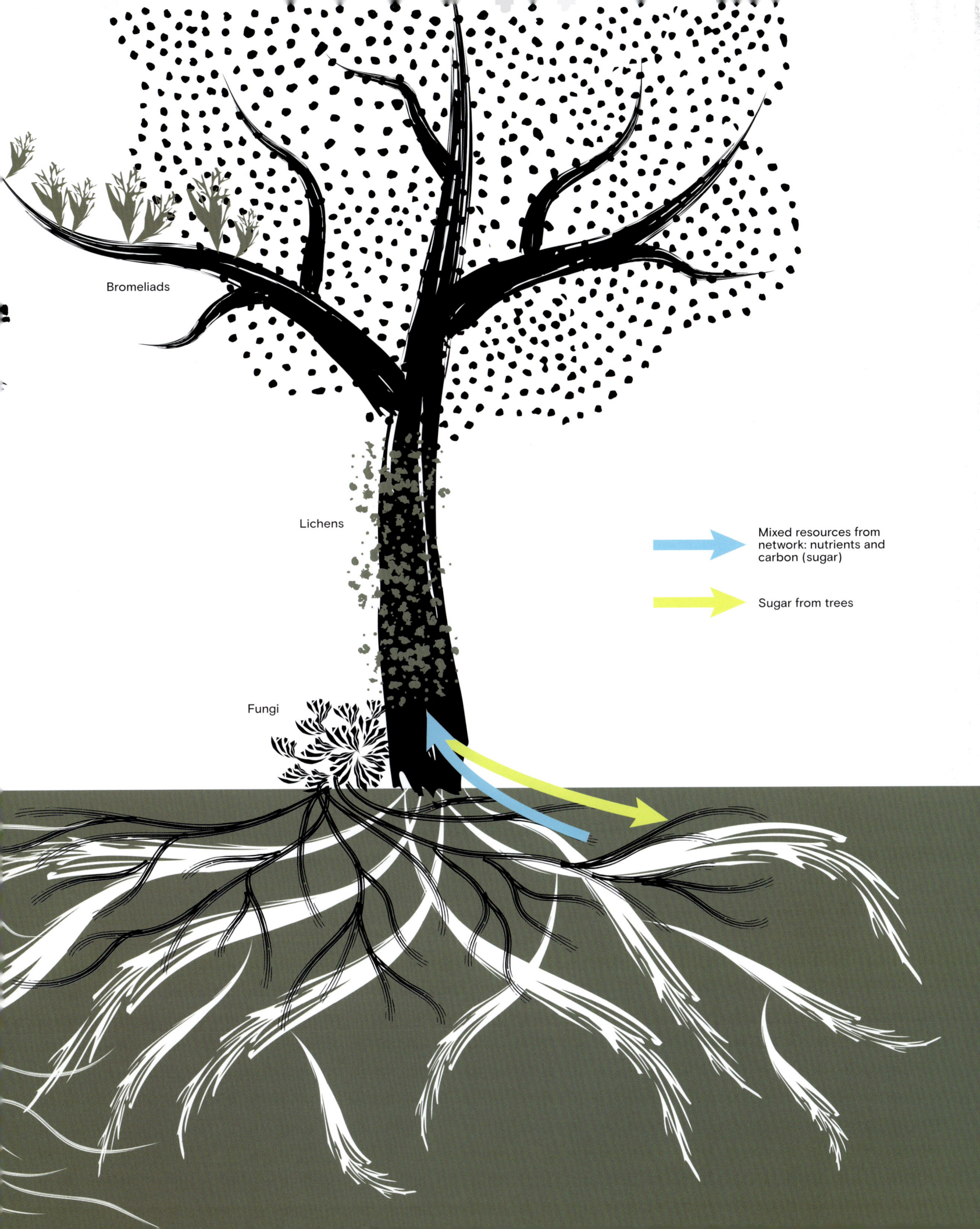

Bromeliads

Lichens

Fungi

Mixed resources from
network: nutrients and
carbon (sugar)

Sugar from trees

Ferns

Ferns (*Polypodiophyta*) are non-flowering vascular plants that reproduce by spores rather than seeds. There are about 10,500 different species of ferns and they are of much older lineage than flowering plants: early forms appeared in the middle Devonian period nearly 400 million years ago. Many ferns are epiphytes and live on trees.

01

02

01 – Ferns
Polypodiophyta
Ferns need water for
sexual reproduction.
For this reason, epiphytic
ferns are found in damp
places on the tree, including
the wet-weather side
of the tree trunk.

02 – Bird's nest fern
Asplenium nidus
The bird's nest fern grows
as an epiphyte in tropical
forests but is also a popular
house plant.

03 – The Fern Paradise: A
Plea for the Culture of Ferns
Francis George Heath, 1878
Ferneries became popular
in the Victorian era.

04 – Ferns
c. 1880
Albumen print. Ferns
were amongst the earliest
subjects of photography,
as far back as the 1850s.

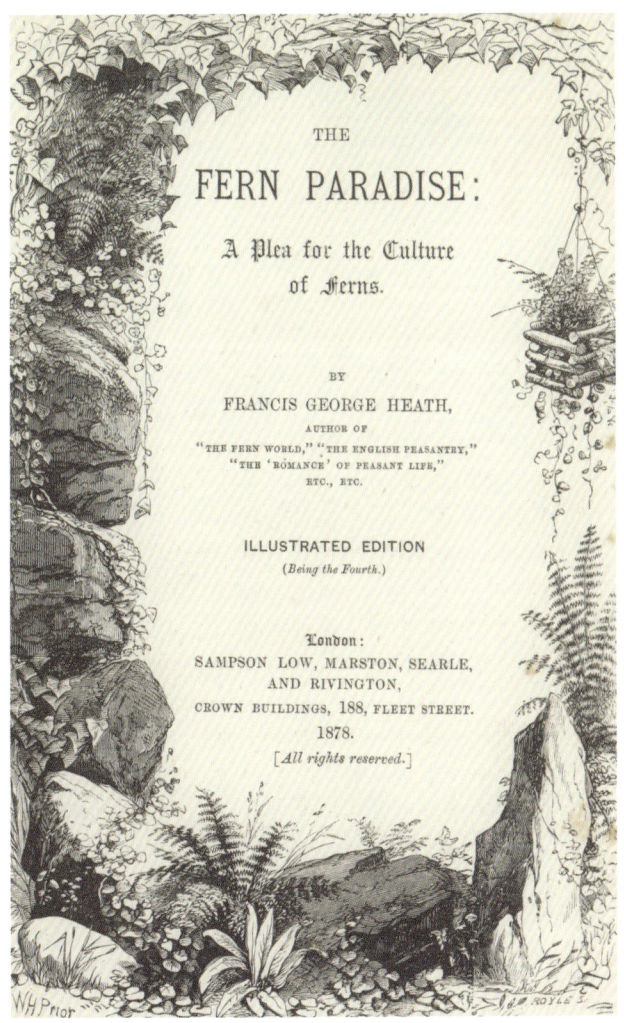

03

04

Lichens and Mosses

Trees are hosts for many other organisms, including lichens and mosses. This means that the loss of a single tree species has a knock-on effect referred to as an 'extinction cascade'. For example, the European ash (*Fraxinus excelsior*) has been shown to host nearly 500 other species, including 87 lichens and 71 mosses.

← Lichen

The mutualistic relationship between the fungus and cyanobacteria or alga that makes up the lichen is obligate, meaning it is essential to both parties. Lichens can also have obligate relationships with trees.

↓ Moss

Mosses are non-flowering, non-vascular plants in the division Bryophyta. There are approximately 12,000 known species of moss, many of which live exclusively on trees.

Threatened Tree Species

There are approximately 17,500 threatened tree species in the world (accounting for about 30 per cent of all tree species). More than 2,000 species are critically endangered and around 440 species are known to have fewer than 50 individuals left in the wild. All the trees listed here are sadly critically endangered.

01

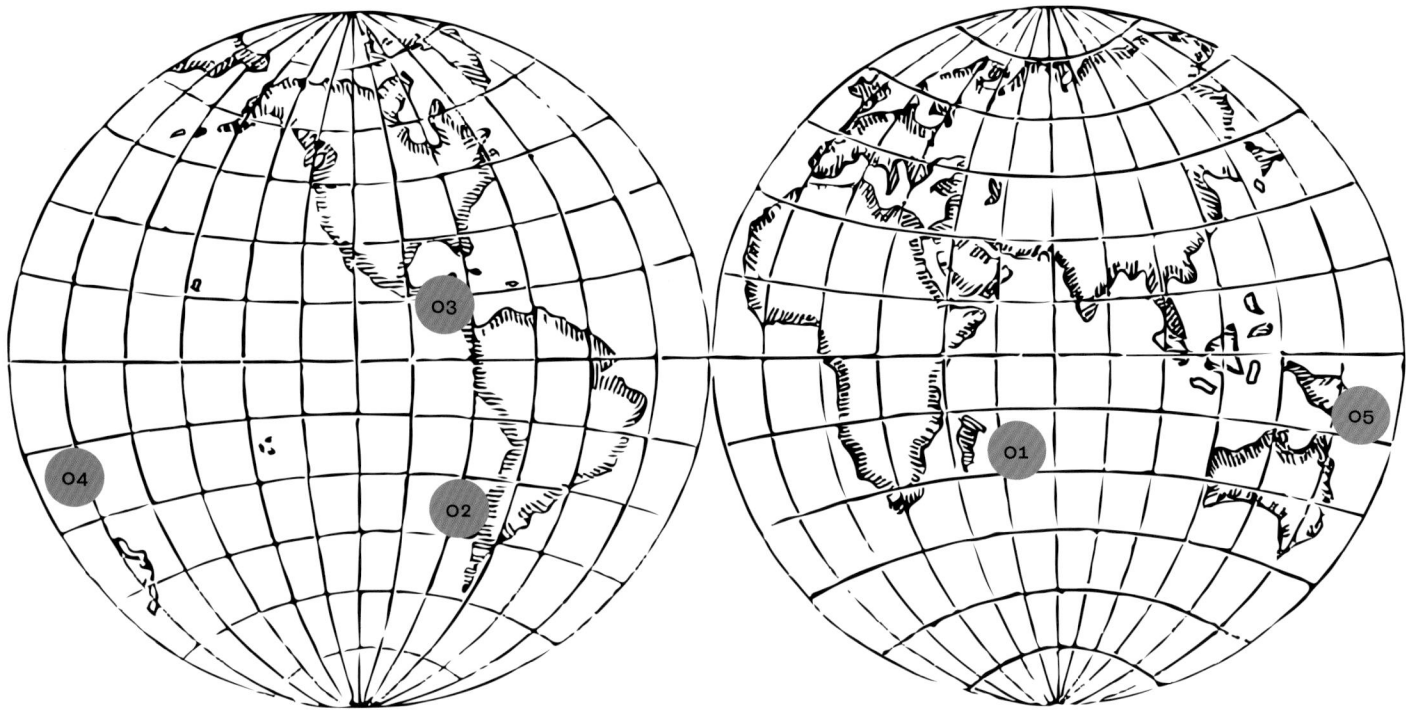

01 – Hyophorbe amaricaulis

Endemic to Mauritius, *Hyophorbe amaricaulis* is extinct in the wild. The last remaining specimen – known as the loneliest tree in the world – resides in its native country in Curepipe Botanical Garden. It is a ripe 150 years old.

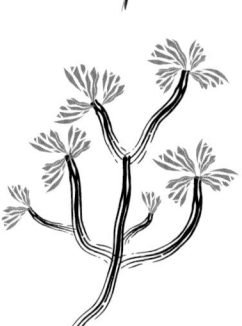

02 – Robinsonia berteroi

On an isle off the coast of Chile, *Robinsonia berteroi* – like its Robinson Crusoe Island home – is named after the eponymous fictional character who was supposedly shipwrecked nearby. It's likely that only one tree of this species remains.

03 – Pleodendron costaricense

Two or three mature specimens of this Costa Rica endemic survive, but with little regeneration evident, they could be the last of their kind.

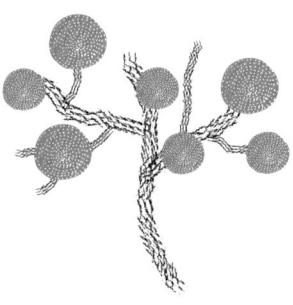

04 – Hibiscus bennetti

Hibiscus bennetti is confined to a single spot on Vanua Levu in Fiji. The area was hit by Cyclone Winston in 2016, after which it is thought that just two specimens remain.

05 – Croton leptanthus

This species is endemic to Lasanag Island, Papua New Guinea, where a solitary individual survives.

Causes of Tree Extinction

By far the greatest cause of tree species extinctions is the clearing of land for agriculture, livestock, urban development and of course logging. Ironically, tree planting for timber and pulp also contributes to tree extinctions.

Clearing forest for agriculture

The single largest cause of tree-species loss is land clearing for agriculture, which accounts for nearly 30 per cent of tree species extinctions. Some trees are affected by more than one threat.

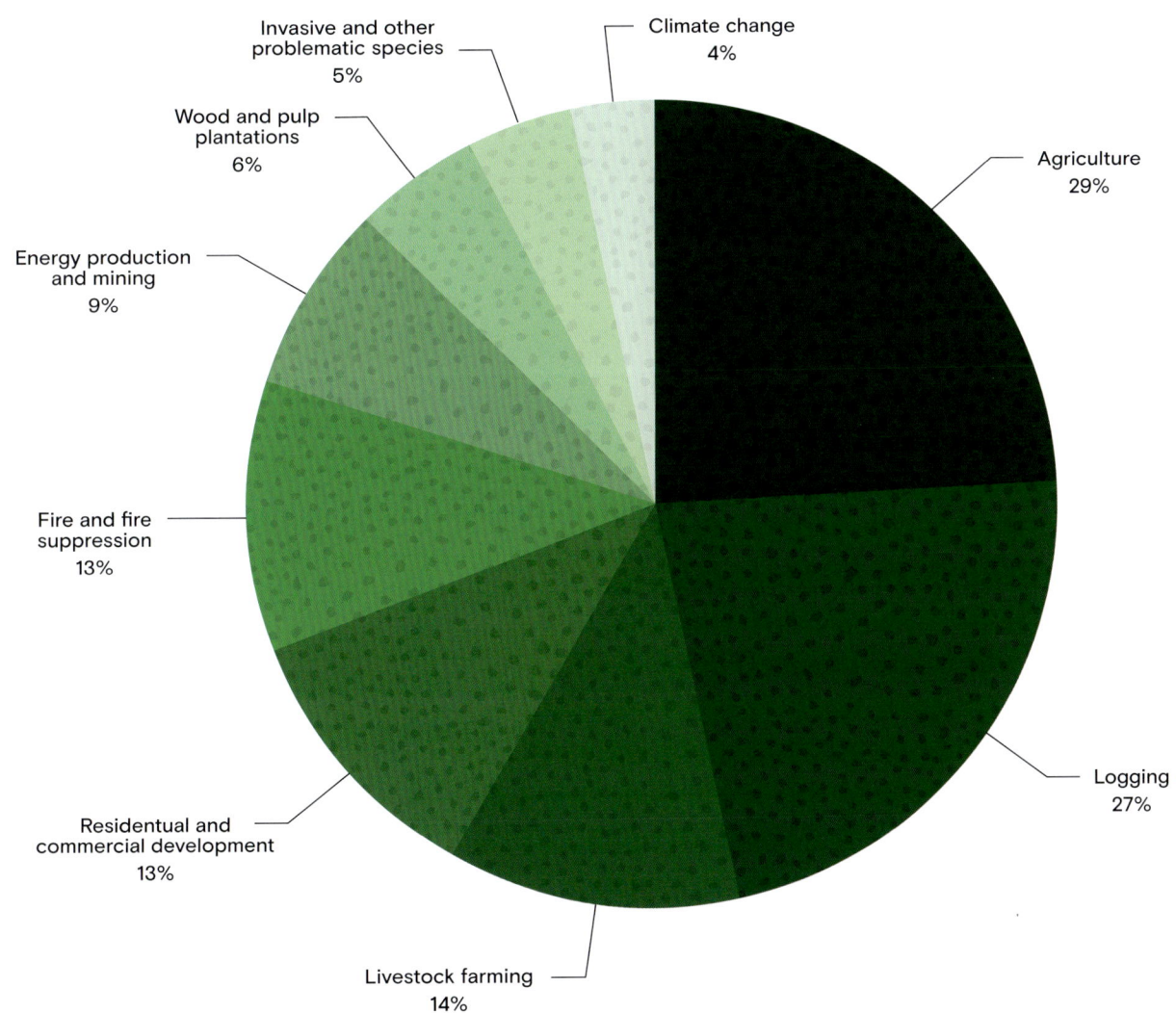

Invasive and other
problematic species
5%

Climate change
4%

Wood and pulp
plantations
6%

Agriculture
29%

Energy production
and mining
9%

Fire and fire
suppression
13%

Logging
27%

Residentual and
commercial development
13%

Livestock farming
14%

Decline in Tree Species

Nearly 30 per cent of tree species are threatened with extinction, with around 140 tree species already known to be extinct. This is undoubtedly an underestimate, as many of the 'data deficient' trees have not been seen for decades.

Forest fire
Fire is thought to account for 13 per cent of tree species extinctions, but the figure is likely to be much higher than this if the setting of deliberate fires to clear forest for agriculture is taken into account.

Extinct
142 (0.2%)

Threatened
17,510 (29.9%)

Possibly threatened
4,099 (7.1%)

Not threatened
24,255 (41.5%)

Data deficient
7,700 (13.2%)

Not evaluated
4,790 (8.2%)

The conservation status of the world's 58,497 tree species

Trees in Children's Literature

Despite our dependence on trees (see Chapter 9, pg.270), our relationship with them is ambiguous. In children's literature, trees are portrayed as both a refuge and something more threatening. Mr Badger's cosy house in *The Wind in the Willows* is often depicted amongst tree roots or even inside a hollow tree trunk, a warm fire burning and Mr Badger comfortably seated in an armchair smoking his pipe. Likewise, in the *Winnie the Pooh* stories, Owl, Piglet and Kanga live in hollowed-out trees with jars of honey and other creature comforts. However, although the Hundred Acre Wood is home to Winnie the Pooh and friends, it is also where the rather frightening Heffalumps lurk. In a similar vein, Australian children in the last century were brought up on the gumnut babies, Snugglepot and Cuddlepie, but the sinister Banksia Men were never too far away. In Tolkien's *Lord of the Rings*, meanwhile, Fangorn Forest is where the Ent, Treebeard, lives; he is described as the oldest creature in Middle Earth. The Ents are animated trees and, although they eventually befriend Pippin and Merry and join the fight against Saruman's forces of evil, there is a sense that these ancient living beings are more elemental than benign.

> In children's literature, trees are portrayed as both a refuge and something more threatening.

For children today, J.K. Rowling's Whomping Willow in the *Harry Potter* stories is a more ambiguous character still. As real *Harry Potter* scholars will know, the Whomping Willow was planted to protect the beloved Remus Lupin – Hogwarts' resident werewolf – perhaps explaining why nobody cuts down an apparently homicidal tree residing in the grounds of a school. In *Harry Potter*, the Forbidden Forest is home to many enchanted (and sometimes dangerous) creatures. It is also where Hagrid hides his giant half-brother, Grawp, where Harry encounters the wounded unicorn and where Firenze the centaur saves Harry's life. Perhaps the most menacing occupants of the Forbidden Forest are Aragog and his family of giant spiders – guaranteed to make arachnophobes quake in their boots.

The *Harry Potter* stories take their cues from ancient European folklore, and the Forbidden Forest very much echoes the woods depicted in older fairy tales, where enchantment and beasts, ancient and old, lurk. However, over recent decades a new narrative for trees has emerged – perhaps best exemplified by the Lorax of Dr Seuss. Published in 1971, when there was growing awareness of environmental issues and the impact of humans on nature, *The Lorax* unashamedly confronts environmental destruction by acting as a champion for the Truffula trees, which are used by the Once-ler to make garments called Thneeds. Not only does the Once-ler fell the Truffula trees to manufacture Thneeds, but his factory pollutes the air and the water, affecting the Swamee-swans and Humming-fish, as well as the small bear-like Barba-loots – who eat Truffula fruits, and consequently become short of food. Eventually the Once-ler cuts down the very last Truffula tree and, with no more raw materials available to make Thneeds, his factory has to close down. The Lorax flies away through the smoggy clouds. Ultimately, the Once-ler realises the error of his ways, handing over the last remaining Truffula seed to a boy and telling him to plant a forest with it.

Today, if you visit a children's bookshop, you will see titles such as *The Giving Tree*, *Wangari's Trees of Peace*, *A Tree is Nice*, *Tap the Magic Tree*, *The Forever Tree* and *We Planted a Tree*. It appears that the narrative is changing for the better, and that children brought up with these stories might do a better job of caring for trees than generations past.

01 – Hansel and Gretel
Liebig vintage trade card, 1896
Hansel and Gretel is a fairy tale related by the Brothers Grimm in 1812 that features a cannibalistic witch who lures two children into the forest with sweets.

02 – Brothers Grimm
German edition cover, pub. Berlin, 1865
Grimm's Fairy Tales, first published in 1812 as *Children's and Household*

Tales, included stories such as *Rapunzel*, *Snow White* and *Hansel and Gretel*. In these tales, the forest is frequently a forbidden and dangerous place.

03 – The Lorax
Dr. Seuss, 1971
Published in the early 70s, The Lorax is one of the first children's books with a positive environmental message about trees, and the iniquity of cutting them down.

01

02

03

Design

Designer Gerardo Osio chose to weave the living branches of a weeping willow tree together to form a 'living seat'. The result, overhanging the Dommel river in Eindhoven, The Netherlands, explores how design can intersect with nature in a responsible and creative way.

'The seat lets the living branches keep growing new leaves and develop more tissue while having enough strength to support a person.'

— *Gerardo Osio*

Trees and Us

ʘ Trees and Us

Many key developments in human history have been marked by our dependence on trees. Strong and durable, their wood has sheltered us for hundreds of thousands of years; today, they continue to provide us with everything from medicine and food to coastal defences and carbon sinks.

01

01 – Carpenter Making a
Chair, Tomb of Rekhmire
c. 1479–1400 BCE
This facsimile copies a wall
painting in the Tomb of
Rekhmire in western Thebes,
Egypt. The detail depicts a
carpenter drilling holes in
the frame of a low chair with
feline legs.

Wood has long been used to build our homes [pg.284], from the construction of rudimentary Neolithic longhouses to elaborate palace structures such as the six-hundred-year-old Forbidden City in Beijing. Similarly, the evolution of wooden ships and wheeled transport enabled the exchange of resources and ideas, facilitating the world's first civilizations to flourish. Timber was also integral to the distribution of the written word. Wooden blocks were used in China to print the world's oldest-known book in 868 CE. Several hundred years later in 1448, the wooden printing press was developed, allowing the mass production of books in Europe for the first time. Printing enabled information to be spread more quickly and at a lower cost. The unique properties of different timber species have resulted in certain species being sought-after for specific uses. For example, the timber of elm trees (*Ulmus*) does not decay when kept permanently wet and was consequently a popular choice for water pipes in medieval England. Similarly, the aromatic wood of the red cedar (*Juniperus virginiana*) is used to line wardrobes due to its ability to deter unwanted clothes moths that attack and damage fabrics.

In addition to timber [pg.164], trees are an incredibly valuable source of nutritious crops, which include fruits [pg.206], nuts, seeds [pg.14], leaves [pg.46], bark [pg.110] and sap [pg.134]. They are particularly crucial for sustaining rural communities between harvests and during extreme weather events such as drought, when other food sources may not be available. Among the forest fruits that have become popular in international markets are avocados (*Persea americana*), lychees (*Litchi chinensis*) and pomegranates (*Punica granatum*). Furthermore, medicine extracted from trees has been of vital significance to human well-being for thousands of years. Their importance is highlighted in *The Canon of Medicine*, composed around 1015, which was an authoritative text for medicine in the Islamic world and medieval Europe for hundreds of years.

The compendium contains the uses of 520 medicinal plants, including 'zarnab' or the common yew (*Taxus baccata*) for setting the heart at ease. Many communities around the world still rely upon trees for traditional medicine, particularly where access to modern pharmaceuticals is restricted. They are also a vital source of compounds to develop prescription drugs for the international market.

Other uses of trees are more indirect, and their true value is often under-appreciated. For instance, they play an important role in reducing soil erosion by binding soil together. In Mali, acacia trees (*Acacia senegal* and *Acacia seyal*) are currently being planted to combat desertification and hold back the Sahara Desert. They also perform an essential function in slowing surface runoff, reducing flooding downstream and increasing local water availability. Trees planted in urban areas can also extract pollutants from the air and improve urban air quality. The World Health Organization estimates that 90 per cent of the globe is subject to either poor or dangerous air quality. Planting appropriate species (for example those with low levels of allergenic pollen) in key areas, such as traffic junctions where air pollution is high, could have a significant impact on improving human wellbeing. Lastly, trees are extremely important for supporting biodiversity and provide a home for many species through the provision of food and habitat [pg.246]. The English oak (*Quercus robur*) alone is estimated to support at least 2,300 species [pg.51].

> The timber of elm trees (*Ulmus*) does not decay when kept permanently wet and was consequently a popular choice for water pipes in medieval England.

Wood-based Products

Trees afford us with many products that are integral to our everyday lives. However, it is their timber that provides the most important direct economic contribution: the total value of global exports of timber products in 2019 was estimated at US$244 billion. Timber from different species can vary considerably in properties such as colour [pg.162], density [pg.160], strength [pg.164] and aroma. Particular species are therefore selectively targeted depending on their end-use, be it construction, furniture, handicrafts or musical instruments.

People have been making wooden furniture [pg.296] for at least thirty thousand years. Neolithic figures have been unearthed depicting people seated on thrones, and the ancient Egyptians developed sophisticated furniture designs. Species that have long been valued for furniture making include rosewood (*Dalbergia*) and ebony (*Diospyros*) trees, which have both been logged intensively, resulting in restrictions on their international trade. Rosewood is particularly sought after in China to make traditional red 'Hongmu' furniture and this demand has made it one of the most trafficked wildlife products by both value and volume.

Mpingo (*Dalbergia melanoxylon*) is a gnarled and extensively branched tree that grows very slowly, not reaching harvestable age for as long as one hundred years. The inky black heartwood of mpingo, which gives it its western name of East African blackwood, is one of the most expensive timbers in the world. It has exceptional durability, making it perfect for wood carving, and a highly attractive finish. It is also valued for making woodwind instruments such as clarinets and oboes. The Mpingo Conservation and Development Initiative is a pioneering sustainable forest-management programme established through collaboration with 41 local communities in rural Tanzania, overseeing 408,500 hectares (1 million acres) of forest and balancing conservation with local livelihoods. These communities are now harvesting mpingo sustainably and have achieved Forest Stewardship Council certification.

Some of the more unusual uses of timber are driven by their unique properties. For example, the Chinese swamp cypress (*Glyptostrobus pensilis*) is the only living species of its genus and is native to southeast China, Laos and Vietnam. Its roots are very light and spongy, making it a perfect material for buoyancy aids. On the Greek island of Crete, meanwhile, the young branches of the Cretan zelkova (*Zelkova abelicea*) are pruned by local shepherds to make 'katsounes' (traditional Cretan walking sticks) due to the tree's flexible, light and durable timber. In South Africa and Namibia, the quiver tree (*Aloidendron dichotomum*) gets its English name from the use of its hollowed tubular branches to make quivers for arrows, a practice employed by the indigenous San people.

Certain species favoured for their timber also have great cultural significance [pg.288]. Tsenden (*Cupressus corneyana*), Bhutan's national tree, is of immense religious importance, making its timber a much sought-after material for the construction and renovation of Bhutanese holy buildings such as 'dzongs' (temples) and monasteries. Alongside naturally occurring stands, these trees have been widely planted close to notable religious buildings. Many of them have now grown to spectacular heights and are associated with significant Buddhist spiritual masters and scholars. In Indonesia, wooden effigies of the dead (known as 'tau-tau') are carved by the Toraja ethnic group in South Sulawesi. These statues are an emblematic representation of the person who has passed away and are usually found near their burial place. Torajans believe that they will only enter 'Poyo' (the spiritual realm) if funeral ceremonies are performed in keeping with their social status. Tau-taus for those of lower status are usually made of bamboo, whereas middle-class tau-tau are sourced from sandalwood (*Santalum*); tau-tau made from the jackfruit tree (*Artocarpus heterophyllus*) are reserved for those of the highest social class.

> The inky black heartwood of mpingo, which gives it its western name of East African blackwood, is one of the most expensive timbers in the world.

02 – East African blackwood
Dalbergia melanoxylon
Larger specimens of East African blackwood are becoming rarer and harder to find due to demand for this highly sought-after wood.

03 – Pepper Harvest and Offering the Fruits to a King
Boucicaut Master,
c. 1410–12
Tempera on vellum. From the *Livre des Merveilles du Monde*.

02

03

Oaks (*Quercus*) are an iconic genus beloved across the world for their often-imposing stature and characteristic acorns. Approximately 430 species of oak are currently known. Predominately found in the Northern Hemisphere, the greatest diversity of species exists in Mexico, China and the United States. Oaks are of immense economic and ecological value; in the Mediterranean, the cork oak (*Quercus suber*) has been integral to local economies for hundreds of years, producing cork predominately for wine-bottle stoppers [pg.132]. This process involves stripping bark from the tree and slowly drying it in the Mediterranean sun, before it's boiled, graded and cut. Because bark can be periodically removed without killing the tree, cork is the ultimate sustainable resource. If it's managed sustainably, cork production can preserve the oak forests that are home to some of Europe's rarest wildflowers. These invaluable forests also reduce soil erosion in an increasingly dry climate.

Food, Medicines and Resin

Products from trees have been an integral part of the human diet for thousands of years. Preserved seeds from archaeological sites have revealed that humans have been collecting and eating wild apples (*Malus*) throughout Europe and West Asia for at least ten thousand years. In South Asia, the mango (*Mangifera indica*) was first domesticated in India more than four thousand years ago, where it was then dispersed throughout the tropics and warm subtropics, becoming one of the world's most indispensable tropical fruits. Many thousands of tree species provide valuable food products and form multi-million-dollar industries.

The remarkable and globally significant fruit and nut forests of Central Asia are home to the living ancestors of domestic apples, pears, walnuts, almonds and other valuable food trees. More than three hundred wild fruit and nut species are found in the region, which has been proposed as one of the world's eight centres of crop origin and domestication. These nutritious and delicious foods are thought to have been spread by people travelling along the Silk Road, the ancient thoroughfare linking Western and Eastern civilizations. These wild varieties of domestic crops may contain genetic resistance to pests and plant diseases that could prove vital to meeting future food-security concerns. Unfortunately, these unique forests are under threat, with forty-four tree species known only in the region now at risk of extinction. After the collapse of the Soviet Union in the 1990s, many of the newly independent countries suffered economic instability, which put additional pressure on the local environment. In Tajikistan, for example, the high cost of imported fuel and coal has caused Tajiks to rely on fuelwood from the forests, resulting in widespread deforestation. Overgrazing has also been identified as a key threat. To address these issues, a number of organizations (including Bioversity International, The Global Trees Campaign and the World Bank) are working with local communities to protect rare tree species, increase natural regeneration and plant seedlings.

Maple syrup [pg.298], often used to liberally coat pancakes and waffles, is produced by tapping maple trees (*Acer*). Indigenous peoples of the Americas were the first to produce maple syrup before early European settlers adopted the practice, traditionally using a batch method to boil sap in large kettles over open fires. The development of modern practices has significantly sped up the process and efficiency. Although a sweet sap can be collected from the majority of *Acer* species, the most commercially important species are the sugar maple (*Acer saccharum*), the black maple (*Acer nigrum*) and the red maple (*Acer rubrum*). The commercial production of maple products in North America is concentrated in the northeastern United States and southeastern Canada. Global demand for maple syrup has been on the rise, with per-capita consumption in the United States growing by 155 per cent over the past 35 years.

It is thought that at least fifty thousand plants are used for medicinal purposes, including many derived from tree species. The Andean fever tree (*Cinchona*) is responsible for producing one of the most important drugs in history. Quinine, an alkaloid extracted from its bark [pg.118], can be used as a treatment against malaria, one of the world's deadliest diseases. Native to Ecuador, Peru and Bolivia and already known to indigenous peoples within its range, the arrival of quinine in Europe pre-empted a scramble to obtain it. In the 19th century, European mortality rates from malaria in overseas expeditions and campaigns were high and the acquisition of quinine was seen as a vital tool for empire expansion. To reduce reliance on exported quinine from South America, fever-tree plantations were successfully established by the British in southern India in the mid-1850s and the

04

CHINCHONA NITIDA TREES.

05

drug distributed to locally stationed soldiers and civil servants. Although its importance as an anti-malarial has now waned with the development of synthetic equivalents, quinine is still enjoyed by many in tonic water, giving the drink its distinctive bitter flavouring.

Some tree species secrete a highly viscous substance composed of hydrocarbons called resin [pg.134]. Agarwood trees (from the genera *Aquilaria* and *Gonystylus*) produce an extremely sought-after resin, also known as agarwood, which is used to develop oil, as well as processed products such as perfumes, incense and medicines. Agarwood is the result of the tree's response to a fungal attack in which a dark, dense resin is formed within its heartwood. Its aromatic properties have been prized throughout history, being noted as a fragrant product in one of the world's oldest texts – the Sanskrit *Vedas*, written in 1400 BCE. With many agarwood-producing species now threatened with extinction as a result of over-exploitation, first-grade agarwood can cost up to US$100,000 per kilogram, making it one of the most expensive natural raw materials in the world. It is estimated that the global market for agarwood is worth US$32 billion. Another resin-producing species is the Socotra dragon's blood tree (*Dracaena cinnabari*), a highly distinctive tree with an umbrella canopy endemic to the island of Socotra, Yemen. The island has been separated from the Arabian Peninsula for 34 million years, resulting in the evolution of the island's remarkable flora: 37 per cent of its plant species are found nowhere else in the world. The dragon's blood tree can live to be thousands of years old and produces resin of deep crimson from which it gets its name. Known locally as 'emzoloh', it has a wide range of uses as medicine, paint and makeup.

In addition to providing food, medicine and resin, trees supply products that have a wide and varied usage. In the Caribbean, nickernuts are the smooth and shiny seeds from shrubby leguminous trees such as *Guilandina bonduc*. Bearing an uncanny resemblance to marble, they are used as counters in Antigua's ancient strategic board game, 'warri'. In Nepal and India, the large dry leaves of the sal tree (*Shorea robusta)* are stitched together to make biodegradable leaf plates and bowls, which can be eaten by livestock after use. They provide an environmentally friendly alternative to single-use plastics and their use locally is increasingly being promoted. During the COVID-19 pandemic, the versatility of tree products was highlighted in the Kurram district of Pakistan, where the fresh leaves of the mazari palm (*Nannorrhops ritchieana*) were used to make handmade facemasks to protect against coronavirus when more conventional materials were not available.

Ecosystem Services

As well as the more tangible benefits that trees provide, they are also a source of vital ecosystem services. A role that is fundamentally important yet completely invisible is their ability to absorb carbon dioxide from the atmosphere. In the process of photosynthesis, they draw in carbon dioxide and convert it to biomass (in the form of roots, trunks, branches, needles and leaves). Wood is a very effective carbon sink – it is almost completely composed of carbon, which is stored within the tree for its lifetime and then takes many years to break down upon its death. The US Forest Service has calculated that America's forests sequester 785 million metric tonnes (772 imperial tons) of carbon a year, which is approximately 16 per cent of US annual emissions (depending on the year). Protecting, regenerating and replanting forest that has been lost is one of our most powerful tools in mitigating the effects of climate change.

Healthy soils support our food systems and represent the largest terrestrial carbon store, while trees and the forests they make up play a crucial role in reducing soil erosion. Tree roots help to bind soil together, providing stability and preventing shallow landslides. Without vegetation cover, soil is exposed to wind and rain, causing it to crumble and erode.

> The dragon's blood tree can live to be thousands of years old and produces resin of deep crimson from which it gets its name.

The whitebark pine (*Pinus albicaulis*) is one of the few pines that can survive in the exposed and windy conditions of the high mountains in northwestern United States and southwestern Canada, where it protects against soil erosion by holding the loose and rocky soil together. Interestingly, it has been found that leaf litter and undergrowth are even more vital than the canopy layer for protection against erosion. Monoculture plantations where the soil has been cleared of vegetation and litter have thus led to serious erosion problems, highlighting the importance of retaining natural forest structure and complexity.

Mangroves grow along sheltered coastlines in the tropics or subtropics and are lifesavers for coastal communities. Mangrove trees live within the reach of the tide, in salty and oxygen-poor soil. Their complex root systems, which extend above and below the waterline, act as a sea wall or buffer, absorbing the energy of the waves. Tropical storms, such as typhoons and hurricanes, can trigger storm surges responsible for devastating coastal flooding, but research suggests that mangrove forests can reduce the death toll from tropical storms by up to two thirds. With climate change predicted to cause rising sea levels and more intense and frequent storm events, planting mangroves could prove a significant nature-based intervention to protect coastal communities from its impacts.

Air pollution poses a serious threat to human health, leading to approximately one in every nine deaths each year. Urban areas are of particular concern, where there is often a high density of people and elevated levels of pollutants. Trees can be planted to reduce urban air pollution by dispersing and trapping toxic particles. In the UK, research has found that British native tree species silver birch (*Betula pendula*), yew (*Taxus baccata*) and elder (*Sambucus nigra*) can capture more than 70 per cent of ultrafine particles emitted from vehicles. Cities around the world are now harnessing planting schemes to reduce air pollution. Hebei province, which encircles China's capital Beijing, is creating a 'green necklace' of vegetation around the city to protect it from the emissions of surrounding factories. Meanwhile in France, Paris is planning to plant trees around four of its historic sites in a bid to improve air quality and adapt to climate change.

Urban trees
Urban trees are aesthetically pleasing, and have been shown to increase the value of neighbourhoods in which they are planted.

Trees also play an important role in urban environments [pg.282] by providing shelter from stifling inner-city temperatures. In cities, predominantly made of concrete and tarmac, a 'heat island effect' is created that can increase average temperatures by several degrees. This might not seem like much, but when you factor in the 'albedo effect' (the absorption of solar radiation on darker-coloured surfaces), pollution and the lack of natural airflow, this additional heat can make city life very uncomfortable. Thermal imaging and temperature measurements have shown that the shade provided by urban trees can more than compensate for this, with temperatures as much as 6 °C (43 °F) lower in a tree's shade than on the open pavement. Climate-change predictions are accelerating the impetus not only to green up cities but to plant trees more generally.

There is also an increasing awareness of the positive association between mental and physical health and access to green spaces. With 68 per cent of the world's population projected to live in urban areas by 2050, prioritizing people's access to trees and nature is of the utmost importance. Many botanic gardens are situated in urban areas and are crucial resources for connecting people to nature. In the 1980s, Japan developed the practice of forest bathing – or 'shinrin-yoku' – as a national health programme to combat its stressed workforce. Shinrin-yoku is the process of appreciating nature through all the senses. Indeed, research shows that the oils ('phytoncides') produced by trees can boost our immune system and that the practice both lowers blood pressure and reduces stress hormones.

We live in an age of megacities, and they don't come much larger than Shanghai in China, which is home to an estimated 28 million people. In cities like this, real estate is hugely expensive, and one result of this is that green spaces are in short supply. An innovative solution to this problem is to bring trees into people's homes: Heatherwick Studio's 1000 Trees building creates an urban forest like no other, the topography of the building complex mimicking an undulating hill upon which – in giant, cup-like receptacles – one thousand trees stand [pg.166].

Urban Planning

Trees planted in urban areas can improve air quality. The World Health Organization estimates that 90 per cent of people are being subjected to either poor or dangerous air quality. Planting appropriate species in key areas, such as traffic junctions where air pollution is high, could have a significant impact on improving human health and well-being.

← Central Park
New York City, USA
Urban green spaces provide cooling, significantly mitigating the 'heat island effect' caused by the reflective heat of concrete and tarmac.

↓ Luxembourg Gardens
Paris, France
City gardens also provide much-needed recreational space to city dwellers, enabling physical exercise, family excursions and mental relief.

Timber Houses

Wood has been used to build our homes for more than 10,000 years, and remains a popular construction material today. Across the world, wood is harnessed to create structures including log cabins, stilt huts and sacred temples.

01

02

03

04

01 – Teahouse
China

Wood has been used as a building material in China for seven millennia, from the grand halls of the Forbidden City to low-slung houses and shopfronts.

02 – Medieval timber frame
England

Timber frames of medieval houses were usually made from hardwood, with infilling panels made from wattle (wooden strips) and daub (clay or plaster).

03 – Gasshō-style house
Japan

Japan's Gasshō-zukuri style is characterized by a thatched, steeply slanting roof. The roof is supported by stout, oak or cedar beams called 'chonabari'.

04 – Tipi
USA

Historically, tipis were used by the indigenous peoples of the Great Plains of North America. They were constructed from wooden poles and animal hides or, in some cases, barkcloth.

05 – Stilt house
Papa New Guinea

Stilt houses are still built and inhabited in Papua New Guinea, particularly along the south coast. The height of the stilts is usually 3.5–4 m (10–12 ft).

06 – Neolithic longhouse
Channel Islands

This replica Neolithic longhouse has been constructed on Jersey using traditional materials and techniques, including mud daubing, thatching and bark-stripping.

07 – Wooden house
USA

Wooden houses are often still preferred over brick or stone in the USA because they are comparatively cheap to build, incur lower taxes, and are safe and warm.

08 – Log cabin
USA

Wooden houses also hark back to European pioneer days, when early settlers built their cabins from the only readily available materials, principally trees.

05

06

07

08

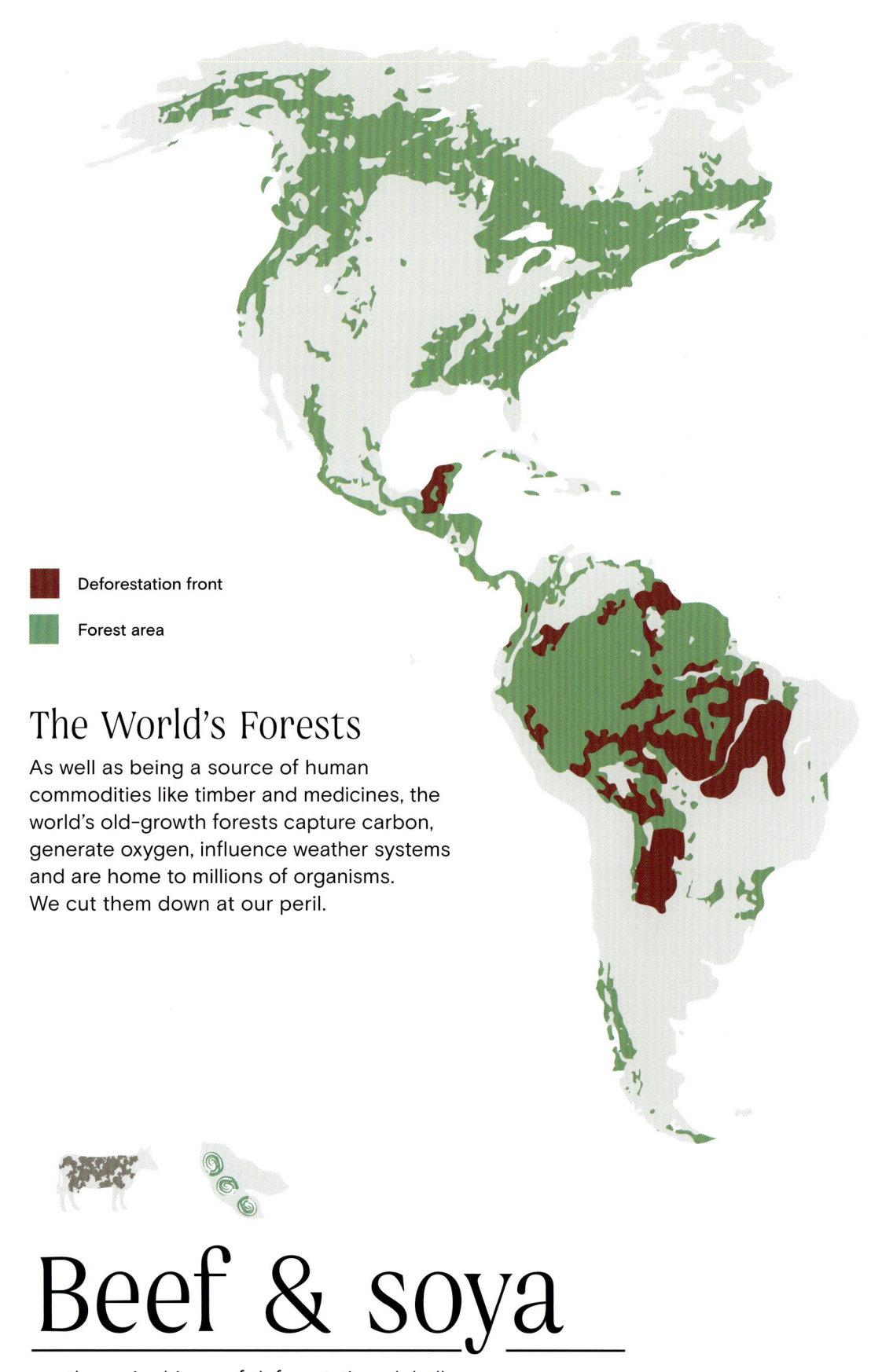

Deforestation front

Forest area

The World's Forests

As well as being a source of human
commodities like timber and medicines, the
world's old-growth forests capture carbon,
generate oxygen, influence weather systems
and are home to millions of organisms.
We cut them down at our peril.

Beef & soya

are the main drivers of deforestation globally

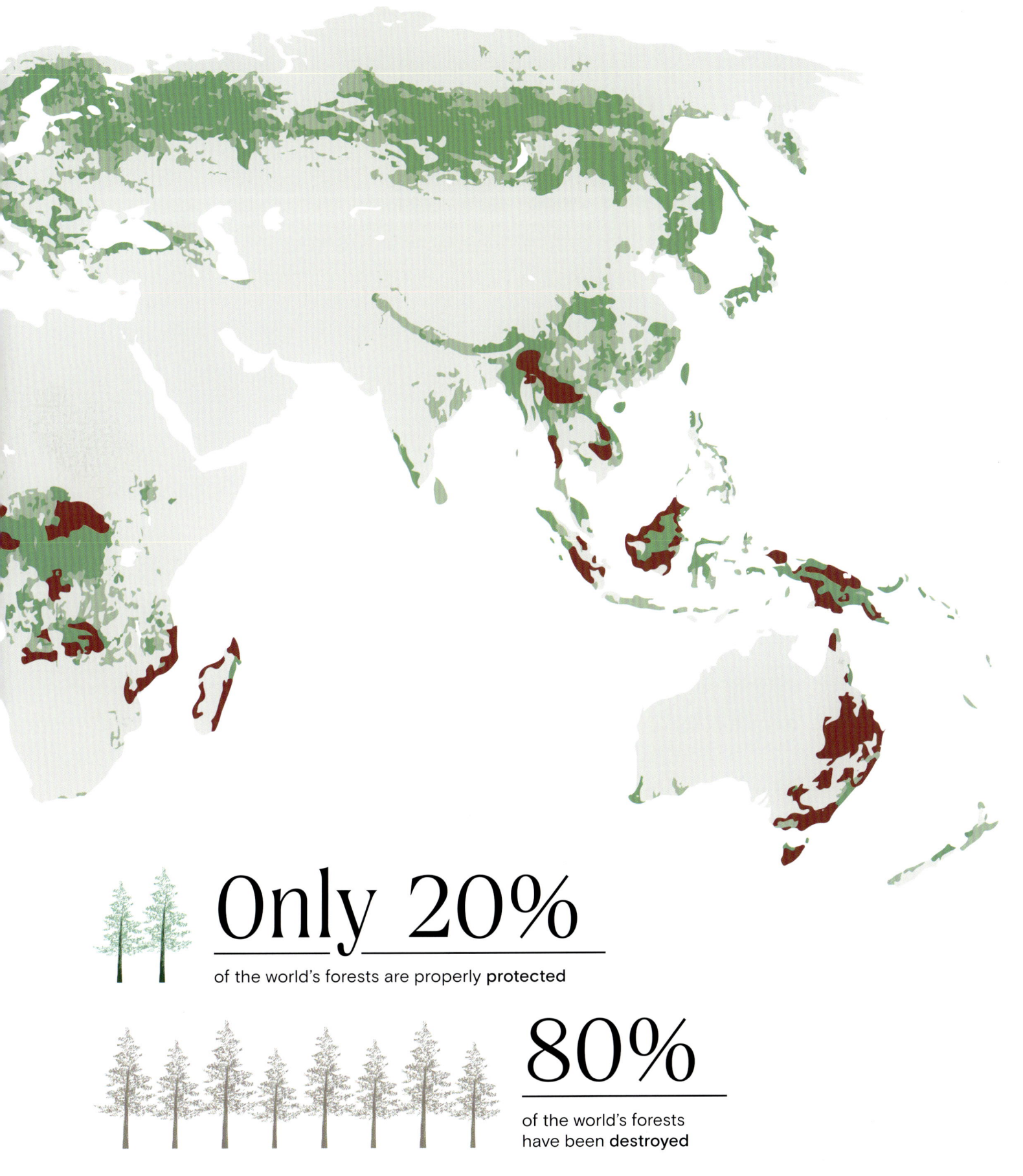

Only 20%

of the world's forests are properly **protected**

80%

of the world's forests
have been **destroyed**

Sacred Timber Buildings and Objects

From churches to monasteries, and from triptych to icons, wood is the material of choice. In many cases, specific trees are regarded as holy, from the yews (*Taxus baccata*) of English churchyards to the tsenden (*Cupressus*) of Bhutanese dzongs. For Christians, the fact that Christ died on a 'tree' has special significance.

↓ **Borgund Stave Church**
Norway
This wooden church was built in the 'stave' style around 1200 CE. The name derives from the building's structure, which is based on weight-bearing posts, called 'stafr' in old Norse.

→ **Sacred objects**
Polychrone wood
Wooden figurines are associated with virtually all religions, one exception being Islam, where they are usually regarded as 'haram' or forbidden.

Borgund Stave Church – Norway, 1180–1250

Kashyapa – Korea, 1700

Maori gable figure (Tekoteko) – New Zealand, 1820s

Raven rattle – Native America, 19th century

Virgin and Child in Majesty – France, c. 1175–1200

Tree-based Products

Even in today's throwaway plastic society, wood remains essential to our everyday lives. It is still the material of choice for furniture, utensils, musical instruments, games and books.

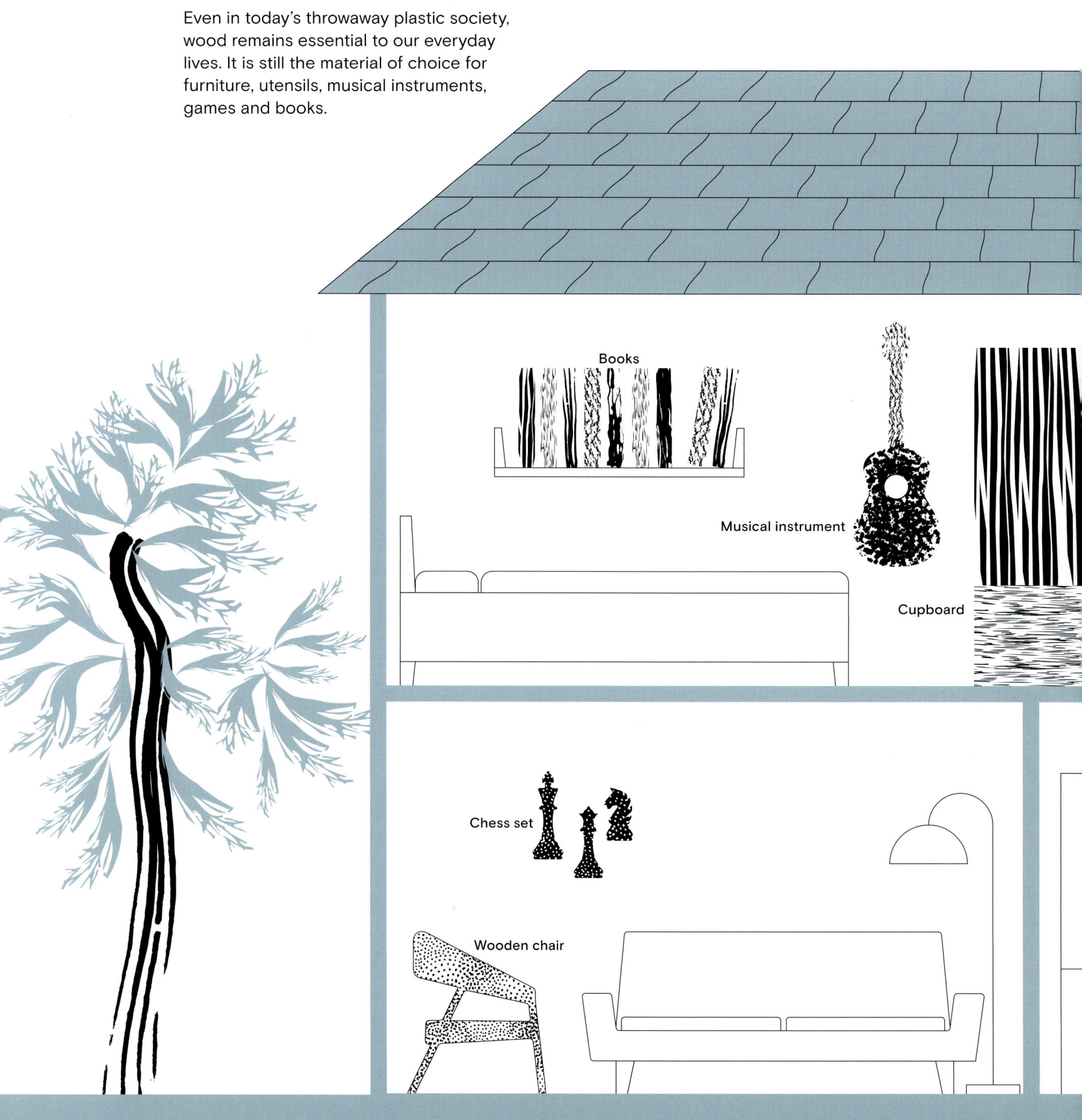

Books

Musical instrument

Cupboard

Chess set

Wooden chair

Medicine

Latex rubber gloves

Chopping board

Chopsticks

Wine cork

Tree fruit

Maple syrup

Table

Ecosystem Services

Ecosystem services can be divided into provisioning services (food, fuel, fibre) regulating services (climate, water), supporting services (soils, nutrient cycling) and cultural services (educational, aesthetic, heritage). Trees provide all of these ecosystem services in abundance, and are therefore central to the ecological functioning of the planet.

Olive tree
Olea europaea
Every part of the tree provides services of some kind to humans and other living organisms on which we depend. Here, baby blackbirds shelter in the trunk of a centennial olive tree.

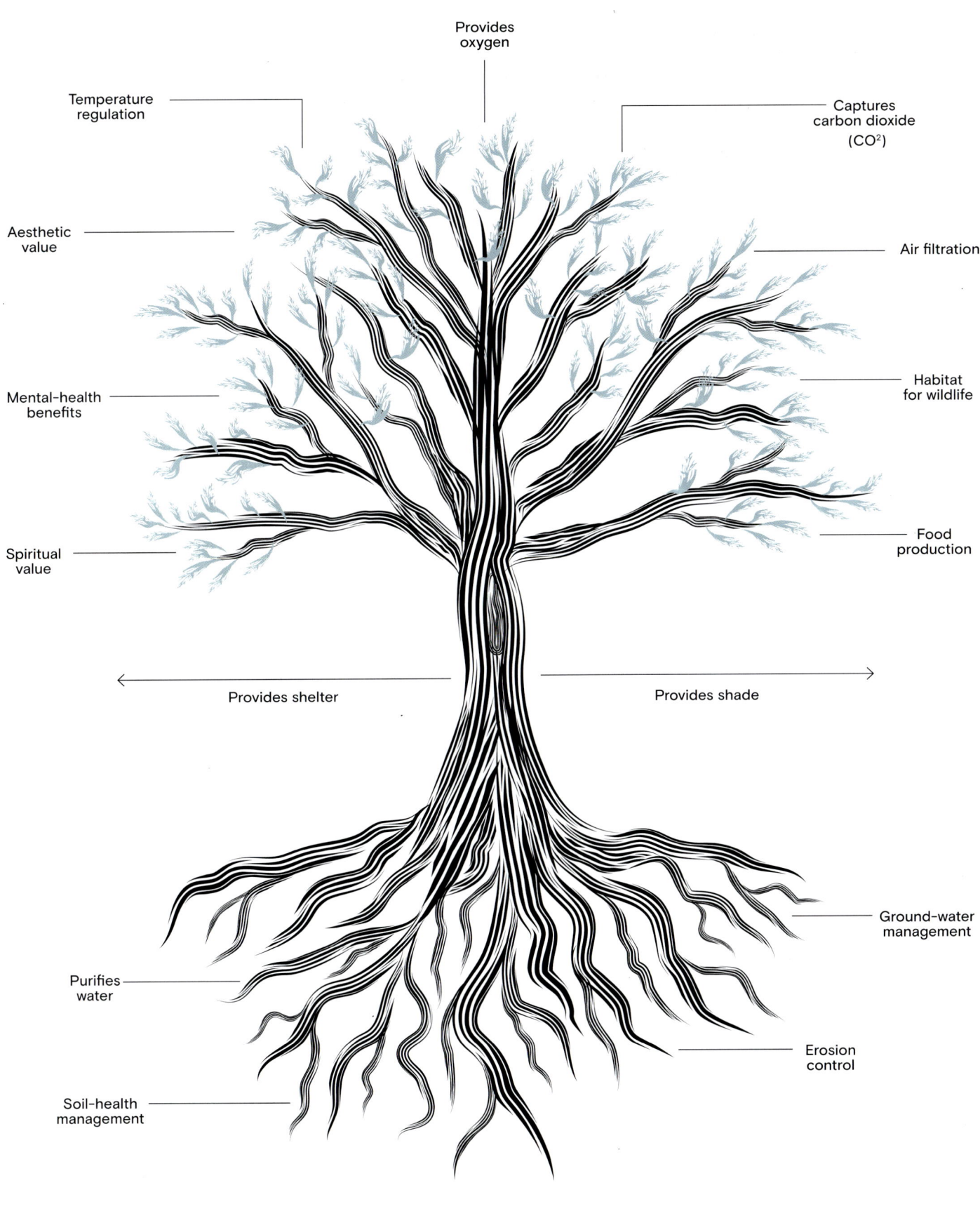

Provides
oxygen

Temperature
regulation

Captures
carbon dioxide
(CO^2)

Aesthetic
value

Air filtration

Mental-health
benefits

Habitat
for wildlife

Spiritual
value

Food
production

← Provides shelter

Provides shade →

Ground-water
management

Purifies
water

Erosion
control

Soil-health
management

The Global Tree Assessment

The Global Tree Assessment was launched in 2015 in recognition of the poor understanding of the conservation status of the world's tree species, which has limited our ability to save threatened species. Various trees are at risk of extinction due to a range of factors: overexploitation for resources, deforestation and the growing dangers posed by climate change and exotic pests or diseases [pg.262]. It is therefore essential to assess the status of the remaining tree species before it is too late. The initiative is being coordinated by Botanic Gardens Conservation International and the IUCN Species Survival Commission Global Tree Specialist Group; it will be used to prioritize conservation efforts to ensure no further tree species go extinct. With nearly 60,000 known tree species, the Global Tree Assessment is the largest biodiversity assessment ever undertaken at a species level.

The Global Tree Assessment was launched in 2015 in recognition of the poor understanding of the conservation status of the world's tree species, which has limited our ability to save threatened species.

01 – Msasa
Brachystegia spiciformis
Msasa occurs from South Africa to Kenya, and is one of the dominant species in African miombo woodland, a habitat covering 250 million hectares (618 million acres).

02 – African redwood
Hagenia abyssinica
This tree species is highly ornamental, with large, attractive flowers. Its male and female flowers range from orange to red to brown.

03 – Grandidier's baobab
Adansonia grandidieri
The statuesque Grandidier's baobab is endemic to Madagascar and comprises Morondava's famous Avenue des Baobabs.

04 – Paraná pine
Araucaria angustifolia
The Paraná pine is found in the Araucaria Moist Forests of southeastern Brazil, Paraguay and northern Argentina. It is Critically Endangered due to unsustainable logging.

01

02

03

04

Furniture

The colours, textures and pliability of wood make it the ideal material for avant-garde furniture design, combining both form and function.

Model 42 low-back
cantilevered armchair
Alvar Aalto, c. 1932
Birch, beech.

01 – Chair of Reniseneb
c. 1450 BCE
Wood, ebony, ivory.

02 – Sgabello
Giuliano da Maiano,
c. 1489–91
Walnut, maple, ebony,
ebonized wood and
fruitwood.

03 – Side chair
Giles Grendey, c. 1735–40
Lacquered and gilded
beech; caning.

04 – Revolving chair
United Society of Believers
in Christ's Second
Appearing, c. 1840–70
Maple, white oak, pine, birch.

05 – Ladderback chair
Charles Rennie
Mackinstosh, 1902
Scottish oak.

06 – Shell lounge chair
Fritz Hansen, 1948
Teak plywood and
stained beech.

01

02

03

04

05

06

Maple Tapping

First made by the indigenous peoples of North America, maple syrup remains a favourite accompaniment to pancakes, waffles and porridge. Today, global production of maple syrup is almost totally confined to Canada and the United States; the Canadian province of Québec contributes around 70 per cent of the world's output all on its own.

01

01 – Sugar house
Vermont, New England, northeast USA

Sugar houses are buildings where sap collected from maple trees is boiled into maple syrup.

02 – Maple syrup industry
Black and white photograph, c. 1930

A brace and bit being used to drill into a maple tree to set up a syrup tap.

03 – Maple syrup advertisement
Lithographic trade card, c. 1880

A Victorian trade card promoting Columbus Avenue Maple Sugar.

04 – Harvesting maple syrup

Maple syrup is best harvested between February and April, when the sap starts to rise. Most trees can produce 20 to 60 litres (5 to 15 US gallons) of sap per season.

02

03

04

Chengdu City

Chengdu is the capital city of Sichuan Province in western China. It is home to 21 million people, and is reported by the Asian Development Bank as China's most desirable city to live in.

← **Green infrastructures**

With 41 per cent urban green coverage downtown and 39 per cent forest cover across the entire municipality, the city of Chengdu has been designated a National Ecological Civilization Demonstration Plot.

↓ Chengdu is prioritizing the construction of ecological zones, greenways, parks, small gardens, and micro-green lands. The goal of the city is to support both economic development and environmental improvement.

Further Reading

BOOKS

David Blakesley, Kate Hardwick and Stephen D. Elliot, *Restoring Tropical Forests: A Practical Guide* (London: Kew Publishing, 2013)
> Based on proven restoration techniques, *Restoring Tropical Forests* enables readers to take practical steps to help save these valuable lands. The book is based on the innovative techniques developed at Chiang Mai University's Forest Restoration Research Unit, Thailand.

Herman Charles Bosman, *The Collected Works of Herman Charles Bosman* (Johannesburg: Human & Rousseau, 1994)
> The life-works of South Africa's best-loved author, Herman Charles Bosman. A writer of extraordinary depth and vision, and a great humourist, Bosman combines a throwaway roughness with mischievousness, selective brilliance and a tragic sense of human fallibility.

Peter Crane, *Ginkgo: The Tree That Time Forgot* (New Haven: Yale University Press, 2013)
> A complete history of ginkgo, the world's most distinctive tree, which has remained stubbornly unchanged for more than 200 million years, providing a living link to the age of dinosaurs.

Nicholas Culpeper, *Culpeper's Complete Herbal: Over 400 Herbs and Their Uses* (London: Arcturus Publishing Ltd, 2009)
> The definitive book on early English herbalism in cookery and medicine.

Jonathan Drori, *Around the World in 80 Trees* (London: Laurence King Publishing, 2018)
> Bestselling author and environmentalist Jonathan Drori follows in the footsteps of Phileas Fogg to tell the stories of eighty magnificent trees from all over the globe.

Carolyn Fry, Sue Seddon and Gail Vines, *The Last Great Plant Hunt: The Story of Kew's Millennium Seed Bank* (London: Kew Publishing, 2011)
> *The Last Great Plant Hunt* tells the story of the work of Kew's Millennium Seed Bank, describing the importance of seed collecting, the process of assemblage and taking care of seeds, the uses of banked seed, and the future of seed conservation worldwide.

Guinness World Records 2022 (London: Guinness World Records Ltd, 2022)
> Celebrating the astonishing, inspirational and eccentric, *Guinness World Records 2022* puts environmental issues front and centre and also encourages readers to make a difference and break records themselves.

Rob Kesseler and Wolfgang Stuppy, *Seeds: Time Capsules of Life* (London: Papadakis, 2014)
> In the compact edition of this lavish collection, a natural history of seeds is presented, illustrated with close-up photographs and scanned electron micrographs.

Richard Mabey, *Flora Britannica* (London: Sinclair-Stevenson, 1996)
> Covering the native and naturalized plants of England, Scotland and Wales, this book is an account of the role of wild plants on social life, arts, custom and landscape.

Desmond Morris, *The Naked Ape: A Zoologist's Study of the Human Animal* (London: Jonathan Cape, 1967)
> *The Naked Ape* has become a benchmark of anthropology and psychology in the second half of the 20th century. When it was first published in 1967, it shocked many people and became the subject of heated debate; the contention was that Morris had written about human beings 'an unusual naked-skinned primate' as though they were just another animal species.

Toby Musgrave, Chris Gardner and Will Musgrave, *The Plant Hunters: Two Hundred Years of Adventure and Discovery Around the World* (London: Cassell & Co., 1999)
> This is the story of the men who travelled through remote and beautiful lands, often at great peril, to collect the plants that have shaped Western garden design for two hundred years.

Toby Musgrave and Will Musgrave, *An Empire of Plants: People and Plants That Changed the World* (London: Cassell & Co., 2000)
> Telling the stories of seven plants whose discovery and cultivation changed the destinies of entire countries, this book investigates the legacy of trade routes overseas and shows how great fortunes were built on espionage, slavery, danger and conflict.

Marianne North, *Vision of Eden: The Life and Work of Marianne North* (London: Royal Botanic Gardens, Kew, 2000)
> Updated edition of the autobiography of the intrepid botanic illustrator Marianne North (1830–90), originally published posthumously in 1893 as *Recollections of a Happy Life*. It is based on her journals of travels in Egypt, Ceylon, India, Japan, Borneo, Australia, South Africa, New Zealand and the United States, and is illustrated throughout by her own surviving watercolour drawings.

Michelle Payne, *Marianne North: A Very Intrepid Painter* (London: Kew Publishing, 2015)
> Marianne North was an English naturalist and a remarkable botanical artist who travelled to more than a dozen countries over the course of more than a decade spent painting the tropical and exotic plants of the world.

Philip Rahtz, *Glastonbury: Myth and Archaeology* (Stroud: Tempus Publishing, 2003)
> Glastonbury, with the distinctive landmark of the Tor, is a familiar name to many. Its fame lies not simply in its renowned festival, but in its legendary associations with King Arthur and with Joseph of Arimathea, whose staff was supposed to have grown into Glastonbury Tor.

Ed. Dennis Reid, *Tom Thomson* (Vancouver/Toronto: Douglas & McIntyre Inc., 2002)
> In this lavishly illustrated, comprehensive and compelling account of Tom Thomson's life and times, six expertly written essays reveal the iconic Canadian artist and colleague of the Group of Seven from many different perspectives, from his biography and work – almost

exclusively landscapes – to the context of the period in which he lived.

David Silcox, *The Group of Seven and Tom Thomson* **(Ontario: Firefly Books Ltd, 2011)**
This award-winning bestseller includes many never-before reproduced paintings and presents the most complete and extensive collection of these landscape artists' works ever published. The four hundred paintings and drawings reveal the remarkable genius of all ten painters who, at some point, were part of the movement.

Paul Smith and Quentin Allen, *Field Guide to the Trees and Shrubs of the Miombo Woodlands* **(London: Royal Botanic Gardens, Kew, 2004)**
Designed for a wide range of people working in, or visiting, south-central Africa, this practical field guide to the trees and shrubs of the miombo woodlands provides an accessible account of sixty of the most common trees and shrubs of the miombo vegetation.

Ed. Paul Smith, *The Book of Seeds* **(London: The Ivy Press, 2018)**
The Book of Seeds takes readers through six hundred of the world's seed species, revealing their extraordinary beauty and rich diversity. Each page pairs a beautifully composed photo of a seed – life-size and, in some cases, enlarged to display fine detail – with a short description, a map showing distribution and information on conservation status.

Wolfgang Stuppy and Rob Kesseler, *Fruit: Edible, Inedible, Incredible* **(London: Papadakis, 2011)**
This book examines why fruits exist and how their short lives are critical to the natural order. Visual artist Rob Kesseler uses scanning electron microscopy to create astonishing images of a variety of fruits and the seeds they contain, while seed morphologist Wolfgang Stuppy explains the formation, development and demise of fruit.

Wolfgang Stuppy, Rob Kesseler and Madeleine Harley, *The Bizarre and Incredible World of Plants* **(London: Papadakis, 2009)**
Illustrated by the striking microphotography of Rob Kessler, with texts by two experts from the Royal Botanic Gardens at Kew, *The Bizarre and Incredible World of Plants* describes the purposes of pollen, seeds and fruit, and the roles they play both in plant reproduction and in preserving the Earth's biodiversity.

Colin Tudge, *The Secret Life of Trees: How They Live and Why They Matter* **(London: Penguin, 2005)**
The Secret Life of Trees explores the hidden role of trees in our everyday lives – and how our future survival depends on them.

Peter Wohlleben, *The Hidden Life of Trees* **(London: HarperCollins, 2017)**
In *The Hidden Life of Trees*, Peter Wohlleben makes the case that the forest is a social network. He draws on ground-breaking scientific discoveries to describe how trees are like human families: tree parents live together with their children, communicate with them, support them as they grow, share nutrients with those who are sick or struggling, and even warn each other of impending dangers.

SCIENTIFIC RESEARCH PAPERS

Daws, M. I., Davies J., Vaes, E., van Gelder, R., & Pritchard, H. W. (2006). Two-hundred-year Seed Survival of Leucospermum and Two Other Woody Species from the Cape Floristic Region, South Africa. *Seed Science Research.* 62: 73-79.

Mitchell, R. J., Bellamy, P. E., Ellis, C. J., Hewison, R. L., Hodgetts, N. G., Iason, G. R., Littlewood, N. A., Newey, S., Stockan, J. A., Taylor, A. F. S. (2019). Oak-associated Biodiversity in the UK (OakEcol). NERC Environmental Information Data Centre. doi. org/10.5285/22b3d41e-7c35-4c51-9e55-0f47bb845202.

Reich, P. B., Uhl, C., Walters, M. B. *et al.* **(1991).** Leaf lifespan as a determinant of leaf structure and function among 23 Amazonian tree species. *Oecologia* **86**, 16–24. doi. org/10.1007/BF00317383.

Schmitt, C., Parola, P., de Haro, L. (2013). Painful Sting After Exposure to Dendrocnide sp: Two Case Reports. *Wilderness and Environmental Medicine.* **24** (4): 471–473. doi:10.1016/j. wem.2013.03.021. PMID 23870765.

WEBSITE RESOURCES

List of Superlative Trees. https://en.wikipedia.org/wiki/ List_of_superlative_trees

Pope Francis (2015). Encyclical Laudato si. https:// cafod.org.uk/content/download/25373/182331/file/ papa-francesco_20150524_enciclica-laudato-si_en.pdf

Seed dispersal. https://en.wikipedia.org/wiki/ Seed_dispersal

Seed Information Database (2008). Royal Botanic Gardens, Kew, London. https://data.kew.org/sid/sidsearch.html

State of the World's Trees (2021). Botanic Gardens Conservation International, Kew, London. https://www. bgci.org/wp/wp-content/uploads/2021/08/FINAL- GTAReportMedRes-1.pdf

Arboreta and Botanic Gardens

The following collection of arboreta and botanic gardens all have interesting and diverse tree collections to visit. The list is not exhaustive but is a good place to start! For a comprehensive selection of arboreta and botanic gardens, search Botanic Garden Conservation International's GardenSearch database, available online at tools.bgci.org/garden_search.php

Algeria: Jardin Botanique du Hamma, Algiers

Argentina: Arboretum Facultad de Agronomia y Zootecnia San Miguel de Tucuman; Arboretum Guaycolec, Formosa; Jardin Botanico y Arboretum 'Carlos Spegazzini' La Plata

Australia: Atherton Rainforest Arboretum and Herbarium Reference Collection, Atherton; Dame Elisabeth Murdoch Arboretum, Cranbourne; National Arboretum Canberra, Canberra; Waite Arboretum, Adelaide

Austria: Alpengarten Franz Mayr-Melnhof, Frohnleiten; Botanischer Garten Innsbruck und Aplengarten Patscherkofel, Innsbruck; Universitat fur Bodenkultur Wien Department fur Integrative Biologie und Biodiversitatsforschung, Vienna

Azerbaijan: Arboretum Azerb NIILH, Barda

Belarus: Arboretum Bel NIILH, Gomel

Belgium: Arboretum Robert Lenoir, Rendeux; Arboretum Wespelaar, Haacht-Wespelaar; Kalmthout Arboretum, Kalmthout

Bosnia and Herzegovina: Arboretum Parsino Brdo, Sarajevo; Arboretum Slatina, Sarajevo

Canada: Montreal Botanical Garden, Montreal; Royal Botanical Gardens, Ontario, Burlington; University of British Columbia Botanical Garden, Vancouver; University of Guelph Arboretum, Guelph

Chile: Arboretum Antumapu, Santiago; Fundacion Jardin Botanico Nacional, Vina del Mar

China: South China Botanical Garden, Guangzhou; Xishuangbanna Tropical Botanical Garden, Xishuangbanna; Shanghai Chenshan Botanical Garden, Shanghai; Kadoorie Farm and Botanical Garden, Hong Kong SAR

Colombia: Jardin Botanico de Cartagena, Turbaco; Jardin Botanico del Pacifico Bahia, Solano

Costa Rica: Arboretum del Bosque Seco Tropical, Liberia; Arboretum Leslie R. Holdridge, San Pedro; Osa Conservation, Puerto Jimenez

Croatia: Arboretum Opeka, Vinica; Botanical Garden of the Faculty of Science Zagreb, Zagreb; Trsteno Arboretum, Trsteno

Czech Republic: Arboretum Kostelec, Czech Agricultural University of Prague, Kostelec; Arboretum Novy Dvur, Steborice; Botanical Gardens and Arboretum, Brno

Democratic Republic of Congo: Jardin Botanique de Kisantu, Inkisi-Kisantu

Denmark: Arboretum Paludosum, Silkeborg; The Greenland Arboretum, Greenland

Estonia: Jarvselja Arboretum, Meeksi

Ethiopia: Forestry Research Centre Arboretum, Addis Ababa; Wondo Genet College Arboretum, Shashemene

Finland: Arboretum Mustila, Elimaki; International Forest Line Arboretum, Turku Botanical Garden 'Botania', Joensuu

France: Jardin Botanique de la Ville de Lyon, Lyon; Jardin Botanique de la Ville et de l'Universite de Tours, Tours; Jardin des Plantes de Paris et Arboretum de Chevreloup, Paris; Jardins des Plantes, Montpellier

Gabon: Arboretum de Sibangu, Libreville

Georgia: Bakuriani Alpine Botanical Garden, Bakuriani

Germany: Arboretum Freiburg-Günterstal, Freiburg; Botanischer Garten München-Nymphenburg, Munich; Späth-Arboretum der Humboldt-Universität zu Berlin, Berlin

Ghana: Bunso Arboretum, Bunso

Honduras: Blue Harbour Tropical Arboretum, Isias de la Bahia

Hungary: Buda Arboretum, Budapest; Folly Arboretum and Winery, Badacsonyörs; Gödöllő Erdészeti Arborétum – Godollo Forestry Arboretum, Gödöllő

India: Auroville Botanical Gardens, Auroville; Forestry Arboretum, Dhaulakuan; Rhododendron Arboretum, Gangtok

Ireland: Fota Arboretum and Gardens, Carrigtwohill; National Botanic Gardens, Kilmacurragh; The John F. Kennedy Arboretum, New Ross

Israel: Arboretum Department of Natural Resources, Bet-dagan

Italy: Arboreto di Arco - Parco Arciducale, Trento; Orto Botanico dell`Università della Tuscia, Viterbo; Orto e Museo Botanico Universita di Pisa, Pisa

Jamaica: National Arboretum Foundation, Kingston

Japan: Kemigawa Arboretum, Chiba; Kobe Municipal Arboretum, Kobe; The Aritaki Arboretum, Koshigaya

Kazakhstan: Arboretum Szczuczinsk, Szczuczinsk

Kenya: African Forest, Elementaita; Brackenhurst Botanic Garden, Limuru; Friends of Nairobi Arboretum, Nairobi

Laos: Pha Tad Ke Botanical Garden, Luang Prabang

Lithuania: Botanical Garden of Vilnius University, Vilnius; Dubrava Arboretum, Vaisvydava

Luxembourg: Arboretum Kirchberg, Luxembourg

Macedonia: City Arboretum Gorica, Ohrid; Arboretum Opeka, Vinica; Arboretum Trubarevo, Skopje

Madagascar: Antsokay Arboretum, Tulear; Parc Botanique et Zoologique de Tsimbazaza, Antananarivo

Malaysia: Sepilok Arboretum, Sandakan; Taman Kiara Arboretum, Kuala Lumpur

Mexico: Jardin Botánico Francisco Javier Clavijero, Xalapa; Jardin Botánico Xochitla, Tepotzotlán Mexico City

Moldova: Arboretum NPO, Kishinev; Botanical Garden Academy of Sciences of Moldova, Chisinau

Montenegro: Arboretum Radigojno, Kolasin

The Netherlands: Arboretum Oudenbosch, Oudenbosch; Belmonte Arboretum, Wageningen; Trompenburg Gardens & Arboretum, Rotterdam

New Zealand: Eastwoodhill Arboretum, Gisborne; Massey University Arboretum and Gardens, Palmerston North; Rotorua Arboretum, Rotorua

Nicaragua: Arboretum Anita Holmann, Managua

Norway: The Arboretum, Bergen; Tromso Botanic Garden, Tromso; University of Oslo Botanical Garden, Oslo

Peru: Arboretum Jenaro Herrera, Requena; Jardin Botánico – Arboretum El Huayo, Iquitos

Philippines: Arboretum of the University of the Philippines, Quezon City; Northwestern University Ecological Park and Botanic Gardens, Laoag City; Siit Arboretum Botanical Garden, Dumaguete

Poland: Kornik Arboretum, Kornik; Arboretum Lesny Bank Genow Kostrzyca, Milkow; Rogow Arboretum of Warsaw University of Life Sciences Rogow

Portugal: Parques de Sintra - Monte da Lua S.A., Sintra

Puerto Rico: Arboretum and Casa Maria Gardens, San German; Arboretum Parque Dona Ines, San Juan

Reunion: Arboretum St. Denis, Reunion

Romania: L'Arboretum Bazos, Timisoara

Russia: Botanical Garden-Institute of the Far Eastern Branch, Russian Academy of Science, Vladivostok; Main Botanical Garden, Russian Academy of Sciences, Moscow; Arboretum Khabarovsk; Arboretum of the Research Institute of Kamyshin

Rwanda: Arboretum de Ruhande, Butare; Bukavu Arboretum/ Garden, Cyangugu

Saint Helena: Clifford Arboretum, St Helena

Sierra Leone: Bo Arboretum, Bo City; Kenema Nursery Arboretum, Kenema

Slovakia: Arborétum Borová hora, Zvolen; Botanická záhrada - Univerzity Pavla Jozefa Šafárika, Košice; Lesnicke arboretum Kysihybel Banska, Stiavnica

South Africa: Kirstenbosch Botanical Garden, Cape Town; Misty Hills Botanical Garden and Arboretum, Muldersdrift; National Botanical Garden, Pretoria

South Korea: Baekdudaegan National Arboretum, Bongwa-Gun; Korea National Arboretum, Pocheon-Si; Chollipo Arboretum Foundation, Sowon-Myeon

Spain: Arboretum-Pinetum Lucus Augusti, Lugo; Arboretum Jardí Botanic Pius Font i Quer Lleida; Lugan Arboretum, Leon; Real Jardín Botánico Juan Carlos I, Alcalá de Henares; Fundacion Sales Xardin Arboretum, Vigo

Sri Lanka: Belipola Arboretum, Bandarawela; Dambulla Arboretum, Dambulla; Royal Botanical Gardens, Peradeniya

Sudan: Soba Arboretum Forest Research Institute, Khartoum

Sweden: Arboretum Alnarpskparken, Alnarp; Arboretum Norr, Umea

Switzerland: Arboretum du vallon de l'Aubonne, Aubonne; Musee et Jardins Botaniques, Cantonaux Lausanne; Botanical Garden of the University of Bern, Bern

Tanzania: Usambaras Arboretum, Usambaras

Thailand: Huay Kaew Arboretum, Chiang Mai

Tunisia: Botanical Garden of Tunis, Ariana

Turkey: Ataturk Arboretum, Istanbul; Aegean University Botanical Garden & Herbarium Research and Application Center, Bornova-Izmir; Nezahat Gokyigit Botanic Garden, Istanbul

Ukraine: Arboretum, Veseli Bokovenki; Botanical Garden of Podolensis, Vinnitsa; Donetsk Botanical Garden, Donetsk; Ukrainian National Forestry University Botanic Garden, Lviv

United Kingdom: Bedgebury Pinetum, Kent; Dawyck Botanic Garden, Bellspool; National Botanic Garden of Wales, Middleton Hall; Oxford University Harcourt Arboretum, Oxford; The Yorkshire Arboretum, York; Westonbirt, The National Arboretum, Wiltshire

United States of America: Arlington National Cemetery Memorial Arboretum, Arlington; Arnold Arboretum of Harvard University, Boston; Bartlett Arboretum & Gardens, Stamford; Bernheim Arboretum and Research Forest, Clermont; Boyce Thompson Arboretum, Tucson; Dawes Arboretum, Newark; Harold L. Lyon Arboretum, Honolulu; Minnesota Landscape Arboretum, Chaska; Morris Arboretum, Philadelphia; Morton Arboretum, Lisle; North Carolina Arboretum, Asheville; U.S. National Arboretum, Washington; UC Davis Arboretum, Davis

Venezuela: Fundacion Jardin Botanico Unellez, Barinas; Instituto Experimental Jardin Botánico Dr Tobias Lasser, Caracas

Vietnam: Bidoup Nuiba Botanic Garden, Dalat

About the Authors

Dr Paul Smith is the Secretary General of Botanic Gardens Conservation International (BGCI). BGCI is the largest plant conservation network in the world, comprising 650 member institutions in 100 countries. BGCI leads the Global Tree Assessment, and recently published the 'State of the World's Trees' report, showing that one third of the world's 60,000 tree species are threatened with extinction. With a career spanning thirty years in conservation, Paul joined BGCI as Secretary General in March 2015. Prior to joining BGCI, Paul was Head of the Royal Botanic Garden at Kew's Millennium Seed Bank (MSB). During his nine years at the helm there, seeds from more than 25,000 plant species were conserved in the MSB and, in 2009, the MSB achieved its first significant milestone, securing seed from 10 per cent of the world's plant species, prioritizing rare, threatened and useful plants. Paul trained as a plant ecologist, and is a specialist in the plants and vegetation of southern Africa. He edited the Ecological Survey of Zambia, is the author of two field guides to the flora of south-central Africa, *The Vegetation Atlas of Madagascar* and *The Book of Seeds*. Paul is a Trustee of the National Botanic Garden of Wales, Co-Chair of IUCN's Plant Conservation Committee and Chair of England's Arboreta Advisory Committee. He is the recipient of the New England Wildflower Society's Medal for Services to International Plant Conservation and the David Fairchild Medal for Plant Exploration.

Yvette Harvey-Brown, author of the chapter Trees and Us, has been working to scale up the conservation of threatened tree species at Botanic Gardens Conservation International for over five years. She is particularly interested in facilitating the use of threatened tree species in ecological restoration.

Robert Macfarlane is the author of books about nature, place, language and people including *Underland*, *Landmarks*, *The Old Ways* and *Mountains of the Mind*. His work has been widely adapted for music, film, television, radio and theatre, and he has also written films including *River* and *Mountain*, both narrated by Willem Dafoe. He is a Fellow of Emmanuel College, University of Cambridge.

Acknowledgments

Paul Smith would like to express his sincere thanks to Yvette Harvey-Brown, who wrote the chapter on Trees and Us, and so generously shared her expertise. I'd also like to thank my colleagues in 'Team Tree' at Botanic Gardens Conservation International, as well as the many people from all over the world involved in the Global Tree Assessment and the Global Trees Campaign. Their passion for trees provided the inspiration for this book. Grateful thanks are also due to the team at Thames & Hudson, especially Helen Fanthorpe, Fleur Jones and Lucas Dietrich. Finally, I'd like to thank my wife, Debs, and my daughter, Zaveri, for patiently listening to me read out extracts of the book and offering their advice and encouragement.

Credits

Glossary

alkaloid:
nitrogenous organic compounds of plant origin that have pronounced physiological actions on humans. They include many drugs (morphine, quinine) and poisons (atropine, strychnine)

anther:
the part of a stamen that produces and contains pollen and is usually borne on a filament

anthocyanin:
a blue, violet or red flavonoid pigment found in plants

arboreal:
tree-dwelling

arboretum:
a botanical garden devoted to trees

aril:
an extra seed covering, typically coloured and hairy or fleshy, e.g. the red fleshy cup around a yew seed

autotrophy:
the ability of an organism to produce its own food using light, water, carbon dioxide or other chemicals

bark:
the tough protective outer sheath of the trunk, branches and twigs of a tree or woody shrub

bast:
fibrous material from a plant, in particular the inner bark of a tree such as the lime, used as fibre in matting or cord

berry:
a simple fleshy fruit that usually has many seeds, such as the grape and tomato. A berry is derived from a single ovary of an individual flower

biodiversity:
the variety of plant and animal life in the world or in a particular habitat

biomass:
the total quantity or weight of organisms in a given area or volume

bonsai:
the cultivation of miniature trees using techniques of root and crown pruning

boreal:
the northern biotic area characterized by the dominance of coniferous forests

browser:
an animal that feeds mainly on the leaves of trees, shrubs and herbs

cambium:
a thin formative layer between the xylem and phloem of most vascular plants that gives rise to new cells and is responsible for secondary growth

chlorophyll:
a green pigment, present in all green plants, that is responsible for the absorption of light to provide energy for photosynthesis

commensalism:
an interaction in which only one party benefits, but in a non-detrimental way to the other

cortex:
tissue of unspecialized cells lying between the epidermis (surface cells) and the vascular, or conducting, tissues of stems and roots

cryopreservation:
the process of cooling and storing cells, tissues or organs at very low temperatures to maintain their viability

cryoprotectant:
a substance that prevents the freezing of tissues, or prevents damage to cells during freezing

cultivar:
a plant variety that has been produced in cultivation by selective breeding

cypsela:
a dry, single-seeded fruit formed from a double ovary of which only one develops into a seed, as in the daisy family

deciduous:
a plant that drops its leaves – usually over the winter period – each year

dehiscent (fruit):
a fruit that splits open

dendrochronology:
the science or technique of dating events, environmental change and archaeological artefacts by using the characteristic patterns of annual growth rings in timber and tree trunks

dioecious:
a plant having the male and female reproductive organs in separate individuals

domatia:
small cavities that shelter and provide an adequate physical microenvironment for arthropods

drupe:
a simple fleshy fruit that usually contains a single seed, such as the cherry, peach or olive. A drupe is derived from a single ovary of an individual flower

ecotype:
a distinct form or race of a plant or animal species occupying a particular habitat

ectomycorrhizal:
where the thread-like filaments of the fungi form an external sheath around the roots of the tree, exchanging water and nutrients through the root surface

embryo (seed):
the young multicellular organism before it emerges from the seed

endemic:
only found in one country

endosperm (seed):
nutritive tissue and regulatory structure that nourishes the developing embryo

endosymbiont:
a bacterium of fungus hosted internally by the tree, for example in root nodules

endozoochory:
seed dispersal via ingestion by vertebrate animals (mostly birds and mammals)

epidermis:
the outermost layer of cells covering the stem, root, leaf, flower, fruit and seed parts of a plant

epiphyte:
a plant that grows on another plant

epizoochory:
seed dispersal by transporting on the outside of animals – usually on the hair of mammals

evergreen:
a plant that retains its leaves throughout the seasons

febrifugal:
fever-relieving

filament:
the stalk-like portion of a stamen, supporting the anther

glochidia:
hair-like spines or short prickles, generally barbed, found on the areoles of cacti in the sub-family *Opuntioideae*

grazer:
an animal that feeds mainly on grass

hardwood:
the wood from a broadleaved tree (such

as oak, ash or beech) as distinguished from that of conifers

haustorium:
a slender projection from the root of a parasitic plant or from the hyphae of a parasitic fungus, enabling the parasite to penetrate the tissues of its host and absorb nutrients from it

heartwood:
the dense inner part of a tree trunk, yielding the hardest timber

hemi-parasite:
a plant that can both photosynthesize its own food but which also parasitizes other plants

herbarium:
a reference collection of pressed and preserved plant specimens

herbivore:
an animal that feeds on plants

hermaphrodite (flower):
flowers that have both female (ovary, style, stigma) and male (stamen) parts

hesperidium:
a fruit with sectioned pulp inside a separable rind, e.g. an orange or grapefruit

hydrophobic:
repels water

hyphae:
the branching filaments that make up the mycelium of a fungus

indehiscent (fruit):
a fruit that doesn't split open

infusion:
an extract prepared by soaking plant parts in liquid

keystone species:
an organism that helps define an entire ecosystem. Without its keystone species, the ecosystem would be dramatically different or cease to exist altogether

leached (soils):
where nutrients are drained away from soil, ash or similar material by the action of percolating liquid, especially rainwater

lignotuber:
a rounded woody growth at or below ground level on some shrubs and trees that grow in areas subject to fire or drought, containing a mass of buds and food reserves

monoculture:
the cultivation of a single crop or tree in a given area

monoecious:
a plant having the male and female reproductive organs in the same individual

mutualism:
a symbiotic relationship in which all parties benefit

mycelia:
microscopic, hair-like filaments produced by fungi

mycorrhizal (fungus):
a fungus that grows in association with the roots of a plant in a symbiotic or mildly parasitic relationship

myrmecophyte:
plants that live in a mutualistic association with a colony of ants

nanotechnology:
technology that deals with dimensions and tolerances of less than 100 nanometres, especially the manipulation of individual atoms and molecules

natural capital:
the world's stocks of natural assets, including geology, soil, air, water and all living things

nectary:
a nectar-secreting glandular organ in a flower (floral) or on a leaf or stem (extrafloral)

orthodox (seed):
desiccation-tolerant seed able to survive for long periods while in a dry state

ovary:
the female organ of a flower, the ovary contains ovules, which develop into seeds upon fertilization. The ovary itself will mature into a fruit, either dry or fleshy, enclosing the seeds

parasitism:
a relationship in which one organism benefits at the expense of another

pathogen:
a bacterium, virus or other microorganism that can cause disease

pepo:
an indehiscent fleshy one-celled many-seeded berry (such as a pumpkin, squash, melon or cucumber)

peptide:
a compound consisting of two or more amino acids linked in a chain

petal:
modified leaves that surround the reproductive parts of flowers. They are often brightly coloured or unusually shaped to attract pollinators

pharmacopeia:
a list of medicinal drugs

phenology:
the study of changes in the timing of seasonal events such as budburst, flowering, seed dormancy, fruit set, etc.

pheromone:
a chemical substance produced and released into the environment by an animal, especially a mammal or an insect, affecting the behaviour or physiology of others of its species

phloem:
the vascular tissue in plants that conducts sugars and other metabolic products downwards from the leaves

phyllode:
a winged leaf stalk that functions as a leaf

physical dormancy (seed):
where the hard seed coat needs to be worn away before water can penetrate the seed and germination can take place

physiological dormancy (seed):
where germination is triggered by temperature (termed 'stratification' or 'vernalization') or by specific chemicals

phytotelmata:
cavities formed by the leaves and stalks of plants that collect water

pistil:
the female reproductive part of a flower comprising the stigma, the style and the ovary

pneumatophore:
an aerial root specialized for gaseous exchange

pod:
a seed case that holds a plant's seeds. In many plants (e.g. peas), seeds grow in groups, nestled within a pod

pome:
a fleshy fruit (such as an apple or pear) consisting of an outer thickened fleshy layer and a central core with seeds enclosed in a capsule

recalcitrant (seed):
desiccation-sensitive seed unable to survive after drying

rootstock:
a plant or plant part united with a scion in grafting

samara:
a winged type of fruit in which a flattened wing of fibrous, papery tissue develops from the ovary wall

saprophyte:
a plant, fungus or microorganism that lives on dead or decaying organic matter

sapwood:
the soft outer layers of recently formed wood between the heartwood and the bark, containing the functioning vascular tissue

scion:
a detached living portion of a plant (such as a bud or shoot) joined to a rootstock in grafting

sepal:
the outer parts of the flower (often green and leaf-like) that enclose a developing bud

softwood:
the wood from a conifer (such as pine, fir or spruce) as distinguished from that of broadleaved trees

spectral signature:
a unique pattern of light wavelengths emitted by a plant

stamen:
the male fertilizing organ of a flower, typically consisting of a pollen-containing anther and a filament

stellate:
star-shaped

stigma:
the top of the pistil is called the stigma, which is a sticky surface receptive to pollen

stipule:
small appendages that are usually found on a branch at the base of the leaf stalk

stomata:
minute pores in the leaf or stem of a plant, forming a slit of variable width that allows movement of gases in and out of the intercellular spaces

stratification:
syn. vernalization

style:
a long, slender stalk that connects the stigma and the ovary

syconium:
inflorescence borne by figs (*Ficus*), formed by an enlarged, fleshy, hollow receptacle with multiple ovaries on the inside surface

symbiosis:
interaction between two different organisms living in close physical association, typically to the advantage of both

syncarpium:
a fleshy multiple fruit, formed from two or more carpels of one flower or the aggregated fruits of several flowers

tannin:
a yellowish or brownish bitter-tasting organic substance present in some galls, barks and other plant tissues, consisting of derivatives of gallic acid

taproot:
a straight tapering root growing vertically downwards and forming the centre from which subsidiary rootlets spring

temperate:
a region or climate characterized by mild temperatures

testa (seed):
seed coat

transpiration:
the process by which plants give off water vapour through their stomata

trichome:
a small hair or other outgrowth from the epidermis of a plant, typically unicellular and glandular

trophic:
the level a specific organism occupies in a food chain

vernalization (seeds):
the exposure of seeds to low temperatures in order to stimulate seed germination

xanthophyll:
a yellow or brown carotenoid plant pigment that causes the autumn colours of leaves

xylem:
the vascular tissue in plants that conducts water and dissolved nutrients upwards from the root and also helps to form the woody element in the stem

Index

Page numbers in *italics* refer to illustrations.

Front and back cover illustrations © Here Design
Endpaper (front): Photosampler/Alamy Stock Photo
Endpaper (back): Vyacheslav Saltayev/Alamy Stock Photo

First published in the United Kingdom in 2022 by
Thames & Hudson Ltd, 181A High Holborn, London WC1V 7QX

Trees: From Root to Leaf © 2022 Thames & Hudson Ltd, London
Text © 2022 Paul Smith
Foreword © 2022 Robert Macfarlane
Trees and Us chapter © 2022 Yvette Harvey-Brown

For a full list of picture credits, see page 310

Designed by Here Design

British Library Cataloguing-in-Publication Data
A catalogue record for this book is available from
the British Library

ISBN   978-0-500-02405-8

Printed and bound in China by C&C Offset Printing Co. Ltd

Be the first to know about our new releases,
exclusive content and author events by visiting
thamesandhudson.com
thamesandhudsonusa.com
thamesandhudson.com.au

The captions for the images on the below pages are as follows:

Page 2: Forest canopy, Upper Bavaria, Germany.
Page 16: Pod mahogany (*Afzelia quanzensis*).
Page 48: The colourful, ornamental leaves of the maple (*Acer*).
Page 62: Ginkgo leaf fossils, dating back to the Permian period, 270 million years ago.
Pages 74–75: Drone image of autumnal leaves.
Page 80: Kokerboom (*Aloidendron dichotoma*), a Namib desert tree.
Page 112: Canary date palm (*Phoenix canariensis*).
Page 144: Pedunculate oak (*Quercus robur*) print by Pierre François Legrand, after Gerard van Spaendonck, 1799–1801.
Page 158al: Engraving of the Swanscombe Man and tribe, who lived approximately 250,000 years ago.
Page 158ar: Nineteenth-century chromolithograph depicting the first canoes on the Seine.
Page 158bl: *Man's First Home: A Primitive Rock Shelter* by Robert Ayton. Colour lithograph.
Page 158bc: Stone Age raft. Nineteenth-century lithograph.
Page 158br: Dog-driven wheel for raising water. Nineteenth-century wood engraving.
Page 159al: A large cannon crushes soldiers under its wheels, representing the American Civil War. Wood engraving after J. Tenniel, 1864.
Page 159ac: The HMS *Victory*, Portsmouth, England. Photocrome, *c.* 1900.
Page 159ar: 3D-print Forust homeware from reclaimed wood by designer Yves Béhar.
Page 159bl: Laying the foundations of the Venetian lagoon. Eighteenth-century illustration by Jan van Grevenbroeck.
Page 159br: Conestoga Wagon by H. Langden Brown, 1938.
Page 176: Tree blossom in spring.
Page 208: Pomegranate (*Punica granatum*).
Page 240: San Martín de los Andes forest, Patagonia, Argentina.
Page 272: Urban greenery in Dubai.
Pages 302–03: Silver birch forest, Liminka, Finland.
Page 304: English oak (*Quercus robur*) in England, UK.